New Technology
at Work

New Technology at Work

Arthur Francis

CLARENDON PRESS · OXFORD

1986

Oxford University Press, Walton Street, Oxford OX2 6DP
Oxford New York Toronto
Delhi Bombay Calcutta Madras Karachi
Petaling Jaya Singapore Hong Kong Tokyo
Nairobi Dar es Salaam Cape Town
Melbourne Auckland
and associated companies in
Beirut Berlin Ibadan Nicosia

Oxford is a trade mark of Oxford University Press

Published in the United States
by Oxford University Press

British Library Cataloguing in Publication Data
Francis, Arthur
New technology at work.
1. Industrial sociology 2. Technology—Social aspects
I. Title
306'.36   HD6955
ISBN 0-19-878016-8
ISBN 0-19-878015-X Pbk

Library of Congress Cataloging in Publication Data
Francis, Arthur.
New technology at work.
1. Labor supply—Effect of technological innovations
on.   2. Technological innovations—Social aspects.
3. Quality of work life.   I. Title.
HD6331.F66 1986   338'.06   86-12705
ISBN 0-19-878016-8
ISBN 0-19-878015-X (pbk.)

Text processed by The Oxford Text System
Printed and bound in
Great Britain by Biddles Ltd,
Guildford and King's Lynn

Preface

This book is about the variety of ways in which work may be organised in the future as the new microelectronics technology transforms products and production processes in factories and offices. The central task of the book is to set out the range of alternatives open to us, to point out a number of the conflicts of interest involved in choosing between those alternatives, and to analyse the consequences of those choices. This new technology is already enriching some lives and impoverishing others. There is no inherent logic within the technology which dictates that either of these things should happen. Social and economic processes are at work and result from individuals making decisions. This book is intended to help all those involved in that decision-making, and those who are in a position to comment on the results of their decisions. It is addressed, therefore, to managers, trade unionists, students, and academics whose work and/or interests bring them into contact with the results of the application of the silicon chip.

I argue that, as this major technological development is increasingly incorporated into yet more products and leads to the adoption of increasingly automated office and factory management and information systems, opportunities are created for the design of more interesting and fulfilling jobs and more satisfactory relationships between people at work. It also creates opportunities for switching forms of organisation away from highly bureaucratic management hierarchies. There could be a trend towards more co-operative undertakings and new opportunities for individuals or small groups to set up on their own account rather than work within a large-scale enterprise. For any of these trends or opportunities to be realised, and for the results to be beneficial, those in a position to make choices require knowledge of the alternatives open to them and need power to exercise their preferences. As knowledge in this situation is power, this book is intended to be an aid to those who are seeking to make new technology work, and to make it work for the common good.

This book has its roots both in a research project funded by the Joint Committee of the Science and Engineering Research Council and the Economic and Social Research Council, and in a number of years of teaching undergraduate and postgraduate engineers at the Imperial College of Science and Technology. Preparation of the manuscript began while I was on study leave at the Management Studies Unit at the University of Adelaide, and I would like to acknowledge with thanks the staff there.

Richard Brown, Karen Legge, and Roderick Martin have all read and commented most helpfully on an earlier draft of the book. Chapter 2 has benefited greatly from conversations with David Shepherd and Sue Cook.

In keeping with its subject-matter of new technology at work, this book was written on a microcomputer and fed directly, on floppy disks, into the Oxford University Press Miles system for photo-typesetting. No apology or acknowledgement needs to be made, therefore, to a secretary of unbelievable patience and skill who transformed execrable manuscript into perfect typescript. Sirius and Wordstar did it all. (Or nearly all. At the copy-editing stage it became clear that technology within the publishing process is unevenly developed. Red ink on paper is the only medium used to transform 'manuscript' into acceptable book form. Thanks must go to Loretta Mullens, therefore, who keyed in a multitude of corrections and 'saras' (search-and-replace instructions) marked up by the copy-editor's pen on the microcomputer print-out).

Other authors seem blessed not only with perfect secretaries but with spouses and children who submit themselves with total selflessness to the authorship's outrageous demands. A theme of this book is the conflicting preferences of the various members of any organisation for particular outcomes and the need for negotiation and a fair distribution of power to achieve acceptable solutions. I hope that my wife Jan and our children will feel that this outcome is just such an acceptable solution, resulting from a correct application of those principles.

ARTHUR FRANCIS

February 1986

Contents

1. New Technology at Work: The Issues   1

2. Effects on Overall Levels of Employment and Occupational Structure   9

3. New Technology and the Experience of Work   38

4. New Technology, Work Organisation, and De-skilling: The Theory   61

5. New Technology, Work Organisation, and De-skilling: The Evidence   79

6. Organisation, Control, and Technology: A New Conceptual Approach   104

7. Organisational Design and New Technology   131

8. New Technology: The Challenge to the Unions   154

9. Implementing New Technology   171

10. Conclusions   197

    References   206

    Index   213

For Ruth, Daniel, and Eleanor

New Technology at Work: The Issues

There is hardly any need to begin a book on new technology with an explanation of why the topic is of such significance. *The Microelectronics Revolution*, an edited collection of papers providing one of the earliest and most comprehensive overviews of the impact of new technology, opens with the editor, Tom Forester, quoting Sir Ieuan Maddock's description of microelectronics as 'The most remarkable technology ever to confront mankind' (Forester 1980, xiii).

The heart of this new technology is, of course, a small chip of silicon, measuring less than 10 mm. square, containing extremely complex electronic circuitry which can be used either as the central processing unit or the memory of a computer. The technological breakthrough allowing primitive versions of such chips to be made came in the 1960s, and it was in 1969 that M. E. Hoff at the Intel Corp. in California developed the idea of a computer on a chip.[1] It was not until the late 1970s, however, that public awareness of the far-reaching implications of this technology developed and the debate about its social implications began. In 1977 both *Science* and *Scientific American* published special issues on the so-called microelectronic revolution[2] and in April 1978 the BBC screened the TV documentary *Now the Chips are Down*. This film was widely acclaimed, became a BBC worldwide bestseller, and was the event that really sparked widespread interest in the new technology in Britain.

The reasons for believing the silicon chip to be the cause of another industrial revolution are to do with its cost, capability, and versatility. The facts on cost alone are staggering. To quote from a definitive article on the technology, written by one of the founders of the Intel Corp. for the 1977 special issue of *Scientific American*,

---

[1] The first full-length feature on the microelectronics revolution to appear in a non-specialist publication appeared in *Fortune* in Nov. 1975. Entitled 'Here Comes the Second Computer Revolution', it was written by Gene Bylinsky, is reprinted in Forester (1980), and contains the Hoff story.

[2] Some articles from these special issues appear in Forester (1980).

Today's microcomputer, at a cost of perhaps $300, has more computing capacity than the first large electronic computer, ENIAC [Electronic Numerical Integrator and Computer]. It is twenty times faster, has a larger memory, is thousands of times more reliable, consumes the power of a light bulb rather than that of a locomotive, occupies 1/30,000 the volume and costs 1/10,000 as much. It is available by mail order or at your local hobby shop.[3]

As a result of these developments in microelectronics, the annual use of computing power increased by about 2000 times in the twenty years between 1960 and 1980.

The ENIAC was a contemporary of the famed Morris Minor car. If automobile technology had advanced as fast as computer technology over the same period then the current equivalent BL car, the Metro, would now cost 45 pence to buy, and run for over a year on a gallon of petrol. Every man, woman, and child in the country would own about 100 of them, but garaging them would be no problem as each car would be little bigger than a matchbox.

It is, however, extremely unlikely that any of us could find a use for a hundred of these cars, even if we could solve the problem of getting into them, and this illustrates a fundamental difference between the car and the computer, and another reason for speaking of a microelectronics revolution. This is because of the microchip's wide variety of applications. As I sit at home, typing this book into a word processor, I can count over thirty devices around me which contain an integrated circuit of some kind. These include all the watches and calculators that each member of the household has (10), the laundry equipment (3), an electric drill with electronic control, two radios, the hi-fi equipment (3), two TVs, and video recorder, and two remote controllers for them, a home computer, word processor, and printer, three electronic alarm clocks, a food mixer, toaster, vacuum cleaner, and three dimmer switches for the lights. Moreover, many of these devices containing microelectronics would themselves have been designed and manufactured using processes incorporating microchips. Were I to own a new BMW car, another ten microcomputers would be at my command, so their advertisements claim.

[3] Robert N. Noyce, 'Microelectronics', *Scientific American*, Special Issue on the Microelectronic Revolution, 1977; reprinted in Forester (1980, 29–41).

Beyond this the microelectronics technology is being used on a massive scale in shops, offices, banks, hospitals, and factories. Within shops there are computer-readable price tags to enable the retailer to exercise control over stock and financial transactions, and there are an increasing number of electronic point of sale (EPOS) terminals. Offices are rapidly being automated. Automated telling machines outside banks—'through the wall banking' as it is picturesquely called—are now a familiar part of the High Street scene. Computers are now widely used in hospitals for running machines performing diagnostic services, and increasingly used for patient and hospital management. Within factories major developments have been the applications of computers to process and production control, the introduction of computer-aided design and draughting (CAD), computer numerically controlled (CNC) machine tools, and robots.

We can observe, then, that the new technology has already had an impact on our lives in three major ways. Firstly, it has led to the improvement of existing products, either in their specification or price, or both. For example, quartz-crystal-based watches are both cheaper and more accurate than those driven by clockwork. Electronic calculators can be produced at such a low cost that manufacturers of slide rules and of electro-mechanical calculators have completely lost their markets.

Secondly, it has allowed the creation of entirely new products and markets. Electronic games, home computers, video recorders, and interactive cable or satellite TV are obvious examples of this.

Thirdly, the new technology is having a major impact at work, for a number of reasons. One is the change in product mix brought about by the changing relative price of products which have had microelectronics incorporated into them. As manufactured goods have become relatively cheaper owing to increased labour productivity in manufacturing, there has been a relative increase in consumer spending on services. Hence employment has been shifting from the factory to offices, leisure-related industries, and so on. The incorporation of microelectronics in products and manufacturing processes is likely to accelerate this trend. Another impact of new technology at work is in the way in which products are manufactured. Those incorporating microchips are made in a markedly different way from the more conventional products they have replaced. As one microchip usually replaces many mechanical parts, the amount of labour needed in assembly is reduced. Some of the remaining assembly tasks are more easily automated. Hence there is less routine manual work to do and the relative proportion of white-collar workers within factories rises. Finally, the new technology is itself incorporated into the production process, in the form of, for example, robots, CNC machine tools, or word processors, and so displaces various types of worker.

Commentators on the social and economic implications of the new technology have examined its impact on us as consumers, as citizens, and as workers. With regard to our role as consumers the questions have been to do with patterns of expenditure, changes in life-style, and, in particular, changes in leisure patterns. The questions regarding us as citizens have been about the likely impact of 'the wired society' on such matters as individual privacy; the distribution of income, wealth, and power within any one society; and relations between developed countries and the Third World.[4] In this book we focus solely on a third set of questions: What impact is the new technology having on people at work?

We have just noted that new technology has a major impact at work in accelerating the shift from manufacturing into services, in reducing the amount of routine work, and in job displacement. This last is perhaps the most frequently discussed aspect of new technology at work. To what extent does it destroy, and create, jobs? The use of robots in manufacturing industry is an often-quoted example of job destruction (or, as we might have said in a more optimistic era, job saving). Others are the use of computers in aiding designers and in controlling production processes. Of even greater impact may be the introduction of word processors in the office and the move eventually to the 'paperless office' through full-scale office automation.

---

[4] For an overview of the issues with regard to the individual as citizen see, for example, Klaus Lenk, 'Information Technology and Society', in Friedrichs and Schaff (1982), and for an interesting treatment of the political issues see Benson and Lloyd (1983).

In a society such as the United Kingdom where over 50 per cent of workers are in the service sector compared with only 30 per cent in manufacturing, new technology in the office may have more consequences than robots in the factory for job loss.

We therefore begin, in Chapter 2, by examining the argument frequently put forward that the new technology is leading to wholesale job destruction and hence a permanently high level of unemployment. The simple form of this argument is that new technology is so enormously labour-saving that we will never again need full employment to provide for all our needs. We show that the argument is much more complex, emphasising that the overall level of employment in an economy is influenced by a number of factors. It is subject partly to the normal economic mechanisms, some of which are under the control of central governments through fiscal and monetary instruments. We argue that levels of unemployment are also influenced by a variety of social and economic processes surrounding the way the new technology is introduced and the way its impact is handled. Some of these are amenable to government action and some are influenced by the behaviour of key groups in the society, including management and trade unions.

A second major debate about the effects of new technology concerns the impact on the skills which will be required of workers in the future. Some argue that the dominant effect of the new technology will be to de-skill the workforce, destroying craft occupations and fragmenting jobs into meaningless elements which can be performed by unskilled operators controlled by large-scale bureaucracies run in the interests of international capital. Others suggest that it will be the automated machinery and systems which will take over all the routine tasks, and thus the impact of the new technology will be to require a more highly educated and trained workforce to perform complex tasks which need a high level of human decision-making skill. Such jobs will give autonomy and variety to the worker. The debate is, in other words, about whether the new technology is likely to lead to proletarianisation or professionalisation of workers. We tackle this question by looking first, in Chapter 3, at a number of case studies of work situations where new technology is being used, taking note of the experiences about effects on skill reported there. We then move on, in Chapter 4, to examine in detail the various arguments put forward about the likely effects of technical change on workers' experience of their jobs. In Chapter 5 the theoretical basis of the de-skilling thesis is examined and challenged.

Not only is there dispute about the long-term effects of new technology on skill level, but there is also, closely bound up with this issue, the question of how jobs of the future will be organised. Will they continue to be concentrated in large companies or will there be a growth in the relative importance of small firms? Virtually all large companies operate through a series of smaller establishments. Will these establishments become larger or smaller? In other words, will workers in the future be increasingly herded together in large factories or offices belonging to even larger corporations or will there be a shift towards work being carried out at a more local level within small groups, perhaps operating as independent companies, partnerships, or co-operatives?

In the past most white-collar workers, and particularly those belonging to 'professional' occupations, expected a long-term stable employment pattern, often with the one employer for a large part of their working life. This expectation has taken a severe knock recently because of the effects of the recession leading employers to declare redundancy schemes, even for white collar staff. The question now being asked is whether one of the effects of the new technology might be to lead employers in general to take an increasingly short-term view about employment, irrespective of an economic recession. The ultimate situation would be that such as Rank Xerox are now operating on a small-scale experimental basis whereby some senior staff work for the company from home on a free-lance sub-contractor basis. If the reason for travelling to work in the past was to get one's hands on the work being done, then in the future, as many people's work will be to deal primarily with information, the information can travel cheaply down the cables to the worker at home. And if skills associated with the new technology are specific to a particular occupation rather than to a firm, why should either employer or employee prefer a long-term contract with one employer rather than a consultancy contract for a particular piece of work? The answers to these questions are not obvious and require one to examine a number of assumptions implicit in the way the questions are asked. Chapter 6 is devoted to this exercise.

In Chapter 7 we address the question of what the internal organisation of the office and factory of the future might look like. For those people who are still employed under relatively conventional employment terms, how will they be managed? Will those organisations operating the new technology continue to operate in the standard bureaucratic form or are we likely to see a continued increase in the use of more flexible, less hierarchical forms of organisation. Will there be more use of peer-group, or clan-type, organisations? By extrapolating from present trends, but also from applying the theoretical knowledge we have about typical organisational behaviour, we can engage in some informed speculation.

Any discussion of the possible impact of new technology must take account of the response from trade unions and others representing occupational groups. Much has been written on this from a range of perspectives. There are those who have written from a managerial perspective and concentrated on techniques for gaining acceptance from workers and their unions for the introduction of new technology. Others have been concerned to examine ways in which trade unions can best develop their own interests and bargain most effectively to protect themselves and to share any advantages accruing from the introduction of new technology. Most writers who have this latter perspective have been concerned, quite naturally, with ways in which unions can maintain the integrity of the occupations they represent, protect jobs, and reap some financial benefit. Some, however, have considered ways in which workers and their unions might get involved in issues of job design and in shaping the direction of future technological developments. Their concern has been to prevent de-skilling and to exert pressure to design jobs which give a higher quality of working life. Some reference will be made, principally in Chapter 8, to these arguments. We also discuss at some length the issue of occupational control—that is, that way of organising work where control over the work process resides in the hands of members of the occupational group rather than in the hands of a management hierarchy. Traditional craft workers, and professional groups such as doctors, operate in this manner. What are the limits on this form of organisation, in whose interests is it, and under what circumstances are claims for occupational control likely to be effective? This leads on to a discussion about the way in which new occupations associated with the new technology are likely to emerge. Whereas in the past technological developments led to the emergence of new occupational groups, often with some measure of occupational control and, in the most recent past, usually making claims to some professional standing (for example, the rise of the various branches of traditional engineering), we argue that in the future it is likely that employers will try to ensure that new jobs do not lead to the creation of new occupations as such, and especially not new occupations which will be allowed any measure of occupational control.

The central argument in the book is that there are a variety of ways of organising work, some of which may be more efficient than others, some of which will be preferred by workers, and others which may be preferred by managers. New technology creates new opportunities and, in some cases, new requirements for the way things are organised.

By using the relatively new and rather powerful transactions cost analytical approach, combined with conventional and radical approaches within organisation theory, and insights from the sociology of work, combined with the results from recent empirical work on the introduction of new technology, the intention is to set out some of the possibilities and choices about the use of new technology at work, indicating the different pay-offs to particular groups, and suggesting that the outcomes in specific cases will depend not just on the technology itself but on the strategy and power of the various interested parties.

2

Effects on Overall Levels of Employment and Occupational Structure

Introduction

The aim of this chapter is to summarise the many arguments currently being canvassed about the effects of high technology on employment. In the first part of the chapter we deal with those arguments concerned with the impact on the overall level of employment. In the second part we begin to look at some of the arguments which have been developed about the possible effects of new technology on the kinds of jobs available in the future.

One view of the overall employment implications of new technology is that it is bringing about the collapse of work. One result of this collapse, if it occurred, might be the creation of a two-class society with a small employed elite and and a large underclass of the long-term unemployed. An alternative might be the new technology bringing us to the leisure society with machines doing all the productive work and people freed to follow their own leisure pursuits. Others argue, however, that new technology need have no long-term effect on employment. In their view the economy has the capacity to adjust to technological changes and maintain full employment, though some government action may be needed in order to aid the adjustment process. Our response to these opposing arguments is to suggest that arguments that new technology is creating very large scale unemployment in the near future are alarmist, based on too simple a set of assumptions. Moreover, they usually lay the blame where it is not warranted. Technological change can affect the overall level of employment but the way it does so is complex and is influenced by many factors, not least by the economic and social policies of the government of the day. Most of this chapter is a discussion of these complexities and of possible policy actions, and our emphasis is on the wide choice which should be available in the future over the pattern and intensity of work which people do.

The remainder of the chapter begins to open up the question of what impact new technology is having on whatever jobs may remain. The arguments are not just about the possible extent of de-skilling. There are questions about the possible effects on particular industrial sectors, especially the relative importance of the service compared to the manufacturing sector. For example, will most of the workforce be short-order chefs in fast-food restaurants in the year 2000 while our manufactured products are being made for us by robots in unmanned factories? What will be the effect on the size of firms: does the microchip signal the end of the giant corporation or will it speed up still further the process of concentration among the giant conglomerate multinationals? What will happen to the individual's pattern of employment? Will we all have long-term careers within one company, or will more of us become self-employed, maybe operating on a free-lance basis from electronic offices in our own homes? What will new technology do to the location of employment around the country and across the world? To these vitally important and profound questions there are no simple answers. In this chapter we can only suggest some of the factors which ought to be considered in beginning to think of answers to them. Later chapters will pursue in more depth some of these issues, particularly those dealing with the type of work people will be doing in the future.

The effect of high technology on the level of employment

The BBC documentary that alerted people in Britain to the implication of the silicon chip (Ed Goldwyn's *Now the Chips are Down* that we mentioned in Chapter 1) ended with the alarming questions:

What will happen to the men in today's jobs? Can we all live on the wealth of automatic factories and the earnings of an elite band of 60,000 software engineers?. If we do not [automate], won't our industry be disadvantaged by the automated industries abroad? And if we do automate, will we be able to cope with the problems of large-scale unemployment?[1]

Tom Stonier, Professor of Technology and Society at the University of Bradford and a well-known commentator on the impact of new technology, claims that

It is highly probable that by early in the next century it will require no more than ten per cent of the labour force to provide us with all our material needs—that is, all the food we eat, all the clothing we wear, all the textiles and furnishings in our houses, the houses themselves, the appliances, the automobiles, and so on.[2]

However, despite the publicity given to the more extravagant claims about the impact of new technology on the level of unemployment, and the popular notion that the silicon chip is a job destroyer, a survey, published in 1979, of some 400 documents on the effect of the new information technologies on employment showed 'how little foundation there is to existing studies, half of which are by pessimists (often with a trade union background) and the other half by optimists (who tend to be on the employers' side)' (Institute for Research on Public Policy 1979). An example of the pessimistic view is the study undertaken for the French Government by two of their senior civil servants (Nora and Minc 1980). Their comprehensive examination of the likely social and economic impact of microelectronics technology concluded that with computerisation there had come an end to the creation of jobs in services and a standstill in the industrial labour force, the only industrial jobs created from now being in small and medium-sized businesses. The Batelle Institute in West Germany, on the other hand, in a report published the same year as the original French version of Nora and Minc (1978), said that 'no employment dump caused by technology seems likely' within the report's time horizon of 1990 (Batelle Institute 1978).

Concern about technologically induced unemployment is not new. In the 1950s there was a lively debate about the employment effects of the automation of mass-production pro-

---

[1] Ed Goldwyn in the script of the BBC2 television documentary *Now the Chips are Down* screened in Mar. 1978. This quote is in the text reproduced in Forester (1980, 301).

[2] Tom Stonier, 'The Impact of Microprocessors on Employment', in Forester (1980, 305).

cesses, and by the mid-1960s the Organisation for Economic Co-operation and Development (OECD) was holding a series of international conferences to discuss the problem. During this period a number of alarming forecasts were made. As early as 1950 Norbert Weiner had predicted that automation would result, within twenty-five years, in a depression that would make that of the 1930s seem like a pleasant joke,[3] and in 1965 an Oxford University professor was quoted as saying at one of the OECD conferences:

There seems to the writer little doubt that unemployment due to automation will grow steadily over the next few decades, perhaps centuries, and in the end it is likely to reach a very high figure, say ninety per cent of the labour force, unless radical changes are made in the present pattern of working.[4]

Radical changes were not made and yet unemployment rates remained more or less constant in single percentage figures throughout OECD countries until the oil price shock induced recession after 1973.

Concern about the employment effects of new technology began to re-emerge at the end of the 1970s. Though many commentators drew attention to the job-displacing tendencies of microelectronics technology it is likely that the reason their gloomy forecasts were paid so much attention was because unemployment had been rising fairly fast in most industrialised countries following the two oil price hikes in the 1970s and the consequent world economic recession.

Economic forecasts in the late 1970s were for unemployment to continue to rise in the 1980s. A book by Clive Jenkins, General Secretary of the white-collar union, the Association of Scientific, Technical, and Managerial Staffs (ASTMS), and the union's research director, Barry Sherman, apocalyptically entitled *The Collapse of Work* (1979), did much to reinforce the view of new technology as a job-killer. However, their most alarming quantitative forecast, of 4.5 million unemployed in Britain by 1990, came from forecasts made by the Cambridge Economic Policy Group and was based on economic factors which had nothing to do with new technology. The Cambridge forecasters had simply observed that up until 1986 there would

[3] Quoted by Philip Sadler in Forester (1980, 293).
[4] Quoted by Philip Sadler in Forester (1980, 291–2).

be a substantial increase in the total number of people wanting work, and not enough new jobs being created, because of a low rate of national economic growth, to match that increase. The forecast growth in numbers seeking employment was due to the increase in numbers of young people coming into the labour market, because of a bulge in the birth rate 16+ years ago, and a substantial increase in the number of married women returning to work. In a separate study which Jenkins and Sherman quote, from the Institute of Manpower Studies at the University of Sussex, the estimate was that for there to be enough jobs created to keep unemployment down to the levels of the mid-1970s there would need to be growth in the gross domestic product (GDP) averaging 3.5 per cent per annum in the UK. Not only would this be a post-war record rate of growth if it were achieved, it would have to occur in the face of a worldwide recession. If one makes the more realistic assumption that the GDP growth rate would fall from 3 per cent to nearly zero over this decade then the figure of 4.5 million unemployed comes out of the Cambridge computer.

Contrary to the Jenkins and Sherman argument a group of academic economists (Stoneman *et al.* 1981), in a study published by the OECD, claimed that 'up to now there is no firm evidence that the current high levels of unemployment are due to technical progress' and that 'if account is taken of the offsetting effects that can be predicted, there will be little change in overall demand [for skills and jobs]'.[5]

Those predictions which have been made about the impact of new technology on employment levels have come from two types of analysis. The one is based at the level of the firm. It estimates the number of jobs lost due to improved productivity through the use of microelectronics and then deducts from these losses an estimate of the jobs gained through increased competitiveness and new markets opened up through the use of microelectronics. A particularly ambitious study of this kind, carried out over an entire geographic region, was that done by Ken Green, Rod Coombs, and K. Holroyd of the Tameside region, near Manchester (Green *et al.* 1980). Their conclusion was that if all the companies in that region introduced whatever new technology was available at the time of the study the net

[5] Quoted in Ruffieux (1981).

loss of jobs might be anything from 3 to 9 per cent of all jobs in the region.

The fundamental weakness of this approach is that it relies to some extent on the researchers' imagination as to what the future may hold, particularly about what new goods and services might emerge, either directly from the new technology or simply because its higher productivity may give consumers greater spending power and so create demands for hitherto undreamed-of goods and services. One is reminded of the early expert, but possibly apocryphal, prediction that the maximum number of computers which could be envisaged as being necessary in the UK might be as much as two. Nevertheless, studies about job losses are useful to the extent that they give some flavour of the rate of change associated with the new technology, and they are helpful for any social and economic planning to cope with these changes.

The second type of analysis is the macro-economic approach, deducing future developments from past quantitative macro-economic observations. This approach uses models to deduce future trends in the volume of employment and unemployment from a comparison of expected improvements in productivity and volume of production. There are a number of serious weaknesses with this approach, among them being the difficulty of establishing a correct connection between technical progress and growth, and the conservative assumptions made about the economic and social behaviour of the various parties involved in any technical change.

A third approach is that of scenario building, the approach adopted by Tom Stonier among others. By extrapolating broad social trends such as the long-run shift in industrialised societies from employment in agriculture and manufacturing to employment in service industries he paints his picture of the future where all our material needs are catered for by 10 per cent of the labour force, and the remainder are employed in the service sector, many of them in education and subsidised by the government.[6] Charles Handy, whose work we describe later in this chapter, is another scenario builder.

Turning away from forecasts and towards theories about the relationship between technical change and employment we

[6] Tom Stonier in Forester (1980, 305–6).

find a number of conflicting views. The two extremes can be expressed in the form of two rhetorical questions. The one, posed by those who believe that new technologies do not, in the long run, have any effect on the overall level of employment, is along the lines of 'What happened to all the agricultural workers who left the land as new technology got applied to farming?' In 1870, they would point out, 47 per cent of the population in the United States were employed on the land. By 1970 this figure had dropped to 4 per cent. Few would have imagined, a century ago, that so many new jobs could have been created, bearing in mind the growth in the country's overall population as well as the migration from the farms to the cities. It would have required an enormously creative imagination to have foreseen the kinds of jobs that the children and grandchildren of those farmworkers would now be engaged in.

Another version of the same question is to ask what became of the ostlers, grooms, and blacksmiths as the horse gave way to the internal combustion engine. They nearly all found alternative employment, eventually. The economic theory underlying this optimistic view of technical change is termed 'compensation theory', and we will discuss this after we have looked at the competing rhetorical question.

This, posed by the Nobel Laureate economist Wassily Leontief, asks, not about the fate of stable-hands, ostlers, and grooms, but about the horses themselves. Where are *they* now? 'To argue that workers displaced by machines should necessarily be able to find work building those machines makes no more sense than to expect the horses displaced by mechanical vehicles to have been employed, directly or indirectly, in various branches of the expanding automotive industry,' says Leontief (1978). This pessimistic view, held by such an eminent economist, has to be given serious consideration.

But which of these two views, the optimistic or the pessimistic, is the more correct? Let us first spell out what the theories underlying each of the two positions are saying. Compensation theory argues that the level of employment in the economy is, in the long run, determined only by the overall level of demand in the economy. If the level of demand is kept constant then savings made by productivity gains through the use of new technology in one sector will feed through into some other sector via lower prices, increased wages, or higher profits to investors. Let us imagine, for the sake of illustration, a company being able to change its production process so that it can produce the same number of widgets but with half the number of people. Let us also assume for simplicity that this change does not cost the company anything: it is the result, let us say, of a brainwave the production director had in his bath that morning. (If the change in the production process did have a capital cost then some at least of the cost of that capital would go to pay the wages and salaries of those in the capital goods industry. This then simply represents a transfer of jobs from the widget-making company to the company supplying the new more productive machinery for widget making.) If the change to the new production process is costless, then halving the number employed to produce the widgets represents a commensurate saving in the wage bill. This saving will then be passed on in a combination of three ways. The widget-making company may increase the amount it pays the remaining workers; it may reduce the price of widgets (or not increase it as fast as prices in general rise); and it may be able to retain some of the saving as increased profits. If the last is the case then those profits will either be retained in the company and reinvested, or they will be passed on to the shareholders as increased dividends. The money saved gets passed on, therefore, to create new jobs in one way or another. That going in the form of higher spending power to the workers themselves allows them to buy more goods or services. Buyers of cheaper widgets now have money for other purchases. Directors of the more profitable firm can invest in more capital equipment or, by increasing the dividend, put more money into the pockets of shareholders. The shareholders can then either spend their increased dividends or invest the money. If they spend then this creates more jobs in the consumer industry and if they invest then jobs are created in the capital goods industry. In every case the money, unless it is stuffed into socks and put under the mattress, gets spent on some good or service and therefore, according to compensation theory, creates a series of new jobs in the industries providing these products. While there are substantial objections to this argument which we put forward below, this approach offers a robust corrective to the popular myth that new technology *per se* is bound to destroy jobs on a grand scale. According to compensation theory the leisure society only arrives when people's preference for extra leisure exceeds their preference for extra income. The test for this comes when someone, or some group, is offered a pay rise (in real terms) and the choice either of working the same number of hours and taking home the extra money or of maintaining their real income but reducing the hours worked. If leisure is preferred to income at this point then they will choose the latter alternative. There appears to be little evidence that as a society we have become so rich that a substantial number of people are at this point. In fact one study (Armstrong 1984) reports that over the last century in Britain each 4 per cent increase in hourly wage rates for males has been followed by a 1 per cent reduction in hours worked per year.

Some would argue, of course, that the reason why people still strongly prefer extra income over extra leisure is because they are manipulated by business interests or the ethos of a capitalist society. Our values are distorted by the power of advertising and the ideology fed to us via the mass media, they would claim. However much truth there is in this view, and it should not be discounted entirely, it is not relevant to the current debate as there is no reason why any such manipulation should be reduced as new technology is introduced. A majority of economists would, therefore, argue that the present high level of unemployment has nothing to do with new technology. Depending on their school of thought they would argue that it is due to the national, and world, recession; an adjustment problem caused by demographic changes bringing a substantial net increase of workers on to the labour market; particular problems within Britain at the moment (the unions, poor management, too much or too little intervention by government, North Sea oil, or whatever); or a failure in either fiscal or monetary management of the economy. Unemployment could be reduced given proper economic management and the appropriate institutional reforms, they would claim.

There are, however, a number of problems with this approach. One, especially relevant for the UK and for economically depressed regions of the country, is that there will certainly be international and regional differences in the rate at which new technology is taken up. If there are relative differences in the extent to which labour is mobile between jobs, in the extent of provision for retraining and in the ability and willingness to engage in retraining, in the provision of entrepreneurs to seize new opportunities, and in any other of the factors which influence the rate at which a region or country can innovate, then those regions/nations which are slow are likely to import unemployment from faster-innovating areas. In Chapter 9 we discuss some of the arguments which have been put forward to suggest that British companies are likely to be slower than their international competitors in taking up new technology based products and production processes. There are, of course, devices available to the government to stimulate innovation, and we discuss some of these below.

A second problem with compensation theory is that it ignores the period during which any changes are taking place—the transitional period. Even if in the long run ostlers and grooms become garage mechanics and assembly-line workers, this may take some time and the social dislocations could be very severe. There are frictions in the markets for capital, labour, and products, and it may take a long time for changes in the pattern of demand to work their way through the system. In the meantime the highest costs of these changes are borne by those made redundant as new technologies replace old. A question for government is the extent to which it should rely simply on market mechanisms to steer through these changes or whether it should use, as do all large firms, planning mechanisms to handle changes of this likely magnitude.

A further problem with compensation theory is its neglect of the effect of unemployment benefit, not a factor needing to be taken into account by the earliest exponents of this theory, writing in the mid-eighteenth century. If we return to the example of our widget-making factory after the introduction of the new production process, with half its workforce put on to the labour market (or, putting it more bluntly, sacked), then many of them are likely to be eligible for unemployment benefit. If the unemployment benefit which they are paid is as much as the savings made from the productivity gains from the new production method, then the savings from the use of the new technology do not re-emerge elsewhere to create new demands and new jobs. The savings are taken up by the government in the form of higher taxes and transferred to the redundant workers. A new equilibrium position is established and the net effect is an increase in leisure in the society, forced onto the displaced workers in the form of unemployment. This is not an argument for cutting unemployment benefit but for government action in the form of reflation, job creation, job subsidies, or whatever.

Leontief, whose concern for the horses faced by technical change we have already noted, advances two further arguments against compensation theory.

New machines, new technology introduced because it cuts production costs can indeed reduce the total demand for labor, that is, for the total number of jobs available in all sectors of the economy taken together at any given price of labor—in other words, at any given wage rate (Leontief 1978, 28).

One is that, as capital increases in productivity across a wide range of jobs, labour cannot compete with it on cost grounds. The other is that so many jobs are likely to need a great deal of capital equipment behind them that there will be a shortage of capital equipment and jobs will therefore not be created quickly enough to maintain full employment.

A weak version of the first argument is that put forward by opponents of the trade union movement claiming that, if wage rates had not been maintained at what they call an artificially high level, the introduction of labour-saving equipment would have been retarded and the number of available jobs increased. Leontief rejects this version, asking

by how much would the wages of telephone operators have to be cut in order to prevent the installation of modern, automatic switching equipment? If the wage rate had fallen, say, by 10% and the total employment had increased, as a result, by 5%, there would still have been a net 5% loss in total labor income. (1978, 29.)

A stronger version of the argument, not developed by Leontief himself but based on his theoretical foundations, is that the high level of productivity of capital may mean that in many instances it will be cheaper to use machines rather than people to perform jobs unless the wage rate is well below subsistence level. This argument is only true, however, at the level of the individual firms. It is fallacious at the level of the economy. Unemployment only arises under these circumstances because the public interest cannot be coped with by private interests operating freely in the market place. If, to return to our widget maker, these devices could be made more cheaply by robots unless the wage rate was, for example, half the level of unemployment benefit, the company, in the absence of any government 'interference' would almost certainly use robots. Nevertheless, because those made redundant by the robots are now unemployed and paid unemployment benefit, the choice to use robots is more costly to the economy as a whole than that of using workers, because the total cost per widget is the cost of buying and operating the robot plus unemployment benefit. It would be cheaper for the government to subsidise the employment of widget makers up to the cost of having that person registered unemployed, rather than the company using automated technology. The conclusion is similar to that reached earlier in our argument about the effect of unemployment benefit on compensation theory.

The implication of this conclusion is that full employment could be maintained if the government operated an elaborate system of job subsidy schemes. However, there is an even stronger version of the argument about high wage costs which implies even more extensive government involvement if full employment were to be regained. The full-strength argument is that there are often substantial costs to employing someone which mean that a company might choose to employ a machine rather than a person even if that person's wages cost them nothing. The costs of employing people, rather than machines, include those of selection and recruitment, training, transport to work, provision of a suitable working environment (clean air, good lighting, moderate temperature, and freedom from hazards), and supervision. To take a trivial example, it is a cost rather than a benefit for me to pay my children to clean my car rather than to take it to the automatic car wash, even if the car wash charges more than my children. The labour-intensive car-cleaning method involves either my supervising them carefully or running the risk of the paintwork being scratched by sponges full of grit because they have been dropped in the gutter, and the car bonnet dented by the smallest child climbing on it to reach the windscreen. The implication of this strong argument is that it may become cheaper in straightforward labour cost terms at the level of the economy to pay some people to stay at home while automated machinery produces the goods and services. Only if the social costs of such a strategy, in terms of mental illness, vandalism, etc., were deemed to be high enough, might it be economically efficient to subsidise firms to take on such labour.

Despite these various arguments suggesting high unemployment from rapid technical change which are based on the foundation laid by him, Leontief has himself, with a colleague (Leontief and Duchin 1983), recently published the results of a survey of the likely effects of new technology on the US economy up to the year 2000 which concludes that, far from there being net job losses, there will be a substantial rise in the aggregate labour force over this period due to the rapid take-up of the technology.

Another major theoretical objection to compensation theory concerns the investment behaviour of firms. This argument is fully summarised by Cooper and Clark (1982) but, put briefly, it is that for full employment in the future businesses need to invest at a certain rate in new equipment. The level of investment depends on business people's expectations of the future level of demand, but it has been demonstrated that there is an inherent instability in the economy. If investors assume that the expected level of demand will be lower than the so-called equilibrium position (the position that yields what economists term 'warranted growth') then the level of savings in the economy will exceed investment, and therefore demand will be lower than expected. Investors will, however, assume that their forecasts of demand were too high (whereas in fact they were too low) and revise their investment rates down still further. Thus demand, and therefore employment, continues to fall. Though a number of economists have made further assumptions about the behaviour of the economy—principally, that firms will substitute labour for capital if there is unemployment (because in those circumstances real wages will fall)—which remove some of the apparent instabilities, others have concluded that the instability still occurs because the actual rate of investment may still not necessarily coincide with the warranted rate of growth for full employment. Nevertheless, Cooper and Clark conclude that government could overcome these instabilities by adjusting interest rates (to stimulate investment directly) or by expanding government expenditure (to raise the overall level of demand), if there were no other constraints upon them. There is thus no technological reason for high levels of unemployment: however, because those actions which government could take to reduce unemployment may be contrary to policies aimed at controlling inflation and policies concerned with maintaining the balance of payments, rapid technological change (as it is likely to change the rate of investment by business) may in practice result in increased unemployment.

There are also those who claim to have identified a cyclical pattern to the way in which technical developments arise. It is suggested that there are long waves, of perhaps fifty to sixty years periodicity, in the pattern of economic development of industrialised societies. An explanation for the phenomenon (sometimes known as Kondratiev waves after the economist who first identified them in 1925) was proposed by Joseph Schumpeter and has since been adopted, extended, and supported by empirical studies by Christopher Freeman and some of his colleagues at the Science Policy Research Unit at Sussex University.[7] The notion is that a major innovation, or set of innovations, at a particular point in time is exploited by the ability and initiative of entrepreneurs to create new opportunities for profit. These in turn attract a swarm of imitators who exploit the new opening with a wave of new investment, generating boom conditions. As the innovations mature there will be a period of stagnation and depression, with consequent high unemployment, if or until a new wave of innovations comes along to compensate. It has been suggested that the three previous waves were associated respectively with the steam engine, the railways, and then electric power and the automobile. Barron and Curnow (1979) have suggested that

[7] This is the Science Policy Research Unit (SPRU) programme of research on technical change and employment (the TEMPO study). The theoretical approach is fully set out in Freeman *et al.* (1982), and an accessible source for early empirical results is Freeman (1984).

electronics will be the technological development associated with the fourth wave, with the first, growth, phase being in the 1950s and 1960s and the second, recessive, phase now being moved into.

It is difficult to evaluate the long-wave argument. The line of reasoning is logical, elegant, plausible, and advanced by highly respected scholars. Nevertheless, attempting to detect a cyclical trend on the basis of just three cycles is a risky enterprise. Moreover, the previous waves occurred during periods when the economy and economic theory were much less highly developed. The question must for the moment remain open as to whether it is inevitable that the economy should move in this cyclical way as it copes with the new microelectronics technology.

Our own conclusion from these various debates about employment effects is that there is no reason why the new technology should inevitably mean a massive reduction in employment. Current high levels of unemployment can be accounted for without recourse to an explanation from technology; consumer appetite for yet more goods and services still appears to be insatiable; and even those economists who advance reasons why new technology might cause unemployment at some point in the future acknowledge that just at present the likelihood is that it will cause labour shortages rather than an overall labour surplus. Nevertheless, the rapid uptake of the technology, with the implications this has for very substantial job displacement in some sectors, even if matched in the longer term by job creation elsewhere, is highly likely to create further problems for the economy and for society to handle. How well this will be done, in the midst of a world recession and current difficulties in most countries with high unemployment, remains to be seen.

Effects of new technology on the nature of work

A second area of concern about the effects of new technology has been centred around the question of what changes are likely to come about in the nature of those jobs that are left. Just as the introduction of spinning and weaving machines at the time of the Industrial Revolution in Britain destroyed craft jobs and created fragmented tasks which were performed by unskilled workers, will the new technology also be used to de-skill work? Alternatively, will the new technology improve the quality of work life by being used to perform routine tasks so that those jobs presently performed by unskilled and semi-skilled workers will be automated out of existence. The introduction of this technology will thus lead, the optimists argue, to an increase in the overall level of skill and knowledge required by the working population.

Prominent among those taking the pessimistic view, but active in opposing what he sees to be an extremely undesirable trend, is Mike Cooley, for many years a development engineer and active trade unionist within Lucas Aerospace and one of the architects of the Alternative Corporate Plan put forward by the shop stewards' committee in that company. He describes his fears for the future in his book, *Architect or Bee?* (Cooley, n.d.):

Computerisation and automation will mean that smaller numbers will be required to run the large corporations. These will be a separate elite from the rest of the community and highly organised on the 'business Union' basis as in the United States. They will be true 'Corporation Men', satisfied economically with company cars, company houses and company medicare schemes. There will be funds for the schooling of the Corporation Man's children, special super-annuation schemes and of course expense accounts for overseas travel, entertaining and other corporation 'responsibilities'.

A large sector of those who remain—the unemployed—will be left to fiddle around with 'community work'. These activities will be deliberately chosen because they yield no economic power. Indeed it will be a sort of therapeutic, do-it-yourself social service.. In practice, this will mean that they will spend their time repairing, cleaning up, modifying and recycling the rubbish which the large corporations are imposing on them. Whilst it is conceivable that some of these jobs will be craft based and thus provide an outlet for some initiative and self activism, the significant reality will be that they have no economic power and no industrial muscle. So a significant proportion of the population seems destined to have no 'real job'. (Cooley, n.d., 85.)

A more measured view is taken in a report from the Council for Science and Society, a group of leading figures mainly from academia and industry. Their conclusion is that:

The problems we shall face will not be wholly new ones, but those with which we are familiar; though perhaps intensified and extended. There is likely, for example, to be a progression away from blue-collar jobs to white-collar or service occupations, but these could well take on more of the character which factory work has had in the past, if the tendency is not resisted. (CSS 1981, 93.)

We examine in the next few chapters what evidence there is to support this particularly bleak view.

Charles Handy, an industrialist turned academic turned management guru, in his stimulating book *The Future of Work* (Handy 1984) elaborates four scenarios of the future, first put forward by Tony Watts (1983). He labels them the Unemployment Society, the Leisure Society, the Employment Society, and the Work Society. Underlying all four scenarios is a pessimistic view of the economy's ability to create jobs at a fast enough rate to match the growth in productivity brought about by new technology. The Unemployment Society is what will come about, he suggests, if we merely follow present trends and if we persist, as a society, with our present views about what constitutes a full-time permanent job. In this scenario present employment and income patterns continue, but at a higher level of unemployment. Those in employment earn a relatively high income, but those out of work will continue to exist at rather low levels of unemployment benefit. We would be even more a country of two societies than is already the case, with the divide being between the employed haves and the unemployed have-nots.

An alternative is that in which government takes much greater action to redistribute income. In the Leisure Society the employed are an elite of highly educated and skilled professionals who work full-time, but the wealth created by the equipment they have designed and operate is dispersed rather widely so that the mass of the population are able to live reasonably well off the products of the automated machinery cared for by the core elite. This is the vision foretold in Kurt Vonnegut's novel, *Player Piano*. Such a society would be based on two extreme cultures and supported by highly divergent ideologies—the elite society valuing work and, presumably, awarding themselves status and power as compensation for their efforts, and the underclass valuing leisure and, accordingly, not accepting as legitimate the rewards the elite class have given themselves. Handy rejects a full-blown version of this vision on the grounds that such a divergent society would be very unstable.

The Employment Society would be the continuation of the pattern that existed in industrialised countries up until the mid-1970s, with virtually all those who sought jobs being able to find them, and the typical lifetime employment totalling the 100,000 hours typically worked until recently. For the reasons given above, Handy believes these days will never return if the economy is left to operate on normal market mechanisms. He rejects the idea that industry ought to keep workers on simply to fulfil a responsibility for maintaining full employment and he also argues that it will be impossible to increase the tax base enough to create employment in labour-intensive services such as health and education.

There are, however, a couple of difficulties with this argument. One is that, if there is going to be a high level of unemployment because productivity is rising faster than output, this means that the cost of goods and services is falling fast and either consumer demand is not increasing fast enough or industry cannot gear up production fast enough to create the demanded extra output. If the former, then consumers have money in their bank accounts or pockets that they do not know what to do with and may therefore not mind increased taxation to pay for schools, hospitals, and the social services. If the latter, then there is potential for demand-led inflation, for which one Government remedy would be to increase both spending and taxation. The other weakness in Handy's argument is that there is every possibility that demand for more education and health care will be expressed in the market place as consumers choose to buy these commodities. Thus quite considerable amounts of labour could be generated by demand for private health and education provision. This is not an outcome that everyone would welcome, and it would bring a series of difficulties in its train as inequalities of income would lead to even greater inequalities in access to health care and education; but it is a plausible outcome, and would increase the likelihood of the continuance of the Employment Society.

The Work Society is Handy's preferred future. In it those who want it have some paid employment, probably on a part-time, possibly job-sharing basis, and not for an entire lifetime. People would spend much of the rest of their time doing other kinds of work, on a voluntary basis.

While advocating the Work Society, Handy is enough of a realist to suggest that the most likely outcome, if his assumption turns out to be correct, is some kind of muddled compromise between all four possibilities, with considerable unemployment, some measure of a divide between the haves in work and the 'leisured' have-lesses without jobs, some attempt at the maintenance of full employment as a policy goal, and an increase in the amount of 'work' done on a 'voluntary' basis.

Jonathan Gershuny, now a Professor at the University of Bath and previously a colleague of Chris Freeman at the Science Policy Research Unit at Sussex University, takes a particularly optimistic view of what work will be like in the future (Gershuny 1978). He rejects any notion of de-skilling and suggests that the new technology will open up many opportunities for people to do meaningful work, either in the formal, employed, sector or the informal voluntary sector.

He notes that industrial societies, as they have become more affluent, have shifted employment first from the primary sector (principally agriculture and the extractive industries) to the secondary sector (manufacturing), and then from the secondary sector into the tertiary (service) sector. We have already noted that in the UK, for example, over 50 per cent of those in work at present are employed in the service sector. Gershuny cautions, though, against simply extrapolating these past trends.

In the past, the more affluent in society spent a high proportion of their income buying services provided by relatively poorly paid workers. It is wrong to assume, claims Gershuny, that as levels of affluence rise more generally, as a result of technological developments, the pattern of consumption of the more recently affluent will mirror that of the the Victorian middle and landed classes. Rather than buying in cheap services (no longer on offer because there are no longer very large numbers of an underclass prepared to offer extremely cheap labour) the newly affluent are buying manufactured goods which perform very similar services mechanically. Automatically controlled gas-fired central heating rather than servant-maintained coal fires, dishwashers and automatic washing machines rather than scullery and laundry maids, and power tools and easily applied decorating materials rather than handymen and 'ragged trousered philanthropists' are classic examples of this switch. Gershuny's conclusion is that as new technology continues to be introduced we are likely to move to a 'dual economy' in which there would still be a formal sector of the economy, where there was much use of high technology, a continuing drive for efficiency, and the production of standardised goods, but this sector would increasingly concentrate on the production of consumer durables—in essence capital goods for use in the home. Final commodities and services would increasingly be produced in the home in the informal sector of the economy. One possible implication of this, suggests Gershuny, is that there would be a substantial reduction in the overall numbers formally employed, with the unemployed making creative use of the opportunities these consumer durables make possible in the home—a more benign version of Cooley's vision described earlier. The alternative implication Gershuny draws is that there could be a reduction in the average number of hours employed in the formal sector, with us all making more use of the our increased leisure by using our Black and Deckers, Moulinexs, and Sinclair products. This scenario is in many ways similar to Charles Handy's Work Society, but, whereas Handy based his vision on his views about what values ought to prevail in post-industrial society, Gershuny's picture is built up from an analysis of economic trends.

A more plausible scenario emerging from Gershuny's analysis is, in my view, one in which many of the final commodities and services will actually be produced by individuals or smallish groups, using the mass-produced standardised home-based capital goods, for sale in the market place. We would see the re-emergence of craft-based production, though such production would rely on high technology and large firms at one remove. The implications of this for employment are that we would still expect to see large-scale manufacturing in developed countries, absorbing substantial amounts of labour (though this labour would be involved in knowledge work rather than metal-bashing); and we would expect to see more labour employed again in small firms producing semi-custom-made products or services using machinery produced by the large companies.

It is interesting to explore the philosophical and political underpinnings of these various types of scenario. By and large, those taking a pessimistic view of the likely impact of new technology are those who have misgivings about the capacity of capitalism to produce outcomes that are fair to all citizens. Those who take the view that capitalism operates to favour those with capital at the expense of those who merely have their labour to sell tend to believe that when new technology is developed and applied in a basically capitalist society the result is likely to be that jobs are degraded and the quality of working life of those in employment suffers. Even the actions taken by governments may be inadequate to defend the position of the weak in society. There are those who fear parallels between the Industrial Revolution and today in the way workers are treated as new technology is introduced. As the Council for Science and Society Report (1981) concludes:

The Report has emphasised the inhumanity and injustice which accompanied the early stages of the Industrial Revolution. These were not new things in England, but they had not previously been justified and defended by the prevailing philosophy. So powerful were the effects of this philosophy that to those who looked down from a higher level in society, the suffering became invisible; or if not invisible, then transparent, and their view was not arrested by it but looked through it at what they took to be economic verities beyond.

We have not yet as a nation fully acknowledged the wrong that was done at that time, and we have not fully rejected it. Until we do so, we are unlikely to recover our energy and self-confidence. And until we do so, fresh shoots of the same philosophy will continually arise from its underground roots. (CSS 1981, 93–4.)

Among those who share this view, some would seek to create change by attempting to influence the dominant values of the society. Handy's work can be seen in this light. Others believe that change is more likely to come from pressure being exerted on decision-makers by those with a vested interest in the likely outcomes of the introduction of new technology. They would emphasise the involvement of trade unions at the place of work and the trade union movement as a whole at the national political level. We discuss these arguments in Chapter 8.

The alternative, optimistic, view is taken by those who are confident that a suitable combination of market forces and judicial government action where market forces fail will provide a satisfactory outcome. Gershuny's analysis seems to fall into this category. The changing costs of capital and labour will result in industrial structural changes creating an acceptable society which has come about through the operation of market forces and not because powerful interests have adopted particular technologies which suit them but are detrimental to the interests of weaker groups in society. Much of the rest of this book will be exploring and evaluating the various arguments put forward in elaboration of this position.

Effects of new technology in specific sectors

In addition to these rather heroic forecasts about the post-information-technology society there have been a number of more detailed studies of the impact of new technology on particular industrial sectors. From these a consensus has emerged, which is probably reliable, that the main areas in which there are likely to be substantial effects due to new technology are in the engineering industry, with the increased use of CAD and computer-aided manufacturing (CAM), and the use of robots; in offices through office automation, in which the use of word processors with or without linkages via networks and data bases to form paperless offices is predicted to bring substantial improvements in productivity, though the effects on skills is at the moment quite unclear; and in the computer industry itself, where there is substantially increased demand for software engineers.

Swords-Isherwood and Senker (1980) provide an authoritative survey of the need for skills in the engineering industry. Their conclusions run counter to any simple proletarianisation thesis about the effects of new technology on skill. They argue that there are likely to be, in the UK at least, substantial shortages of electronics engineers, manufacturing engineers, technicians, craftsmen, and people with sales expertise. The low status of manufacturing in Britain, something the UK shares with some other European countries and the United States according to Swords-Isherwood and Senker, has exacerbated the shortage of engineers. To some extent, that shortage is being met by using consultants, particularly by firms in industries which graduates appear to find particularly unappealing, such as motor vehicles and mechanical engineering, but the use of consultants is fraught with difficulty.

They conclude that technicians are likely to be in increased demand for test and inspection work as more and more industries incorporate sophisticated electronics in their equipment. However, the demand for people to do routine testing is likely to decrease as this task becomes increasingly automated, through the use of test equipment incorporating microelectronics.

Craft workers are likely to be in increased demand, at least over the next several years, to carry out maintenance on advanced equipment such as CNC machine tools and robots, though craft work is likely to change its context. We discuss in Chapters 5 and 9 the experiences of BL with reorganising maintenance work when they installed an automated welding line for the Metro car. Their experience well illustrates Swords-Isherwood and Senker's argument that the need will be for increased numbers of such workers, and that they should possess multiple skills: for example, electronics, electrical, and mechanical. In the very long term, automated machinery is likely to incorporate automatic diagnostic equipment in order to simplify maintenance, but it is not obvious that this will necessarily reduce the skill content of craft maintenance work. Already some automated office equipment is able automatically to phone the contract maintenance firm when it develops a fault. The intention is to design equipment so that when the maintenance engineer then calls he, or she, should be able to read off from a display on the machine which circuit board has failed and then simply slip in a replacement. Because of the relatively low cost of such boards, the faulty one would just be dumped. This might possibly be an accurate forecast of the future, but anyone with experience of even a relatively simple up-to-date photocopier will know that the amount of maintenance support it requires goes far beyond this board-changing routine. In the far more complex factory environment the skilled maintenance engineer is likely to have a role for some considerable time yet.

An area of particular weakness in British industry is in the sales and marketing function. Swords-Isherwood and Senker conclude that industry continues to underestimate the significance of marketing and sales functions, and that this weakness, if it persists, will make companies even more vulnerable in the future as the incorporation of microelectronics into products makes these products more complex. The increased need will be for 'salesmen to have a deep technical knowledge of the products they are selling, as well as knowledge of the customers' applications'.[8] Given the low base from which much UK industry starts, this implies substantial job generation for technical experts who can develop competence in selling and marketing.

There have been a number of surveys which have generated specific estimates of job-loss implications of particular technologies. It now seems generally accepted that, for example, one CNC machining centre will replace three conventional machine tools. Thus two machinists' jobs are lost, but CNC machine tools as they are presently used require support from white-collar workers who write the programs which drive them and from maintenance workers to keep them running. Estimating that this totals half a person per machining centre, then the introduction of such machines halves the direct labour requirement. Job-displacement estimates for CAD equipment are harder to establish. A figure given common currency in the engineering industry is that the use of this equipment increases productivity by 300 per cent. The straightforward application of this figure implies that a drawing office could get rid of two-thirds of its draughtsmen if it installed a CAD system. I have heard of no report of this happening. Estimates of actual

[8] Swords-Isherwood and Senker (1980, 174).

job reductions seem to run at about 30 per cent. There appear to be two main reasons for this factor of ten difference between estimate and practice. One is that the figure of 300 per cent increase in productivity just happens to be that which would have had to occur to cost-justify the installation of CAD given capital and wage costs in the early 1980s in the UK when this figure gained currency. The other is that draughtsmen do not spend all their time actually drawing. If, as has been gue-stimated, a draughtsman spends two-thirds of his time doing conceptual thinking, looking up tables, doing calculations, checking, etc., and only one-third of his time with a pencil in his hand at the drawing-board, then an increase in productivity at the drawing-board itself of 300 per cent would mean an overall productivity increase in line with observed job reductions. An implication of this, it should be noted, is that if firms need to justify the use of CAD on cost grounds then they cannot, at mid-1980s UK prices, give each draughtsman a work-station. To utilise fully a CAD system requires only one work-station for three draughtsmen. Each person then has to plan their work carefully and co-ordinate use of the equipment with the others. This clearly represents a reduction in the auton-omy of each worker and, to this extent, a reduction in the quality of their experience of work.

Estimates of the job-displacing effects of robots are highly subject to movements in the capital cost of that equipment and in local labour costs. In 1978 and 1980 two independent German studies[9] reported that, in German firms surveyed, robots had replaced workers at the rate of between two and four men per robot per shift. The exact figures depended on whether the robot was used for material handling or for op-erating a tool, the former application displacing more workers than the latter. It is hard to forecast trends here. Though it is likely that the capital cost of a robot will continue to fall quite rapidly relative to wage costs, there are only a limited number of applications for robots of the current generation of capa-bility. Most assembly work is still too complex to be done by the present generation of relatively senseless machines. Only when touch sensitivity and vision are added to them will they

[9] Battelle Institute (1978) and the West German research institute SOFI—findings reported in an unpublished conference paper.

become capable of much wider application, but then, of course, their cost will be higher.

There have been few studies of office automation applications, but projections about specific skill requirements are extremely difficult to make at present because it is as yet quite unclear how office automation will be used. Wainwright and Francis (1984), on the basis of four case studies, discuss some of the options. Will secretaries, with powerful information technology literally at their fingertips, take over some of the tasks now done by managers or other office-based professional staff; or will managers, once they have a keyboard, have less need for secretarial staff? Will new information technology jobs be created which will be staffed by those with a yet-to-be-designed occupational training? It will be several years before these questions can be answered.

The difficulty with all these forecasts is that the type of skills required depends on decisions made by managers about the way in which work is organised around the technology, and we thus come back to the problem identified with forecasts made about the impact of new technology on society. If managerial decisions are made on the basis of some logic implicit in the technology itself, or in its cost structure, then forecasts can be made with some confidence. If, however, as we have already suggested, and shall be arguing at greater length below, choices about technology and how it is applied are the subject of a conflict of interests, and thus result in bargaining, then predictions about the outcome are more uncertain.

Impact on size of firms and stability of employment

The argument put forward above implies employment in a range of sizes of firm. Now let us turn briefly to the implications of work done by Birch (1979) in the USA and the Science Policy Research Unit in Britain on changes in the size of firms. Birch's work on where jobs were being generated in the US economy stirred up considerable interest around the time of the 1979 election in Britain when Margaret Thatcher's Conservative government first came to power. It was widely misinterpreted as suggesting that small firms were particularly important in the process of job generation. This was one of the more powerful myths leading to the present government's keen interest in policies geared towards the encouragement of small firms. It has now become clear that large firms have remained extremely important in generating jobs and that policies to encourage job creation by large firms are more important, in terms of numbers of jobs created, than those helping small firms. What *is* true is that large firms also destroy large numbers of jobs at much the same rate as they create them, whereas small firms have a lower rate of job destruction. Thus there has been a *net* gain in jobs from the small-firm sector, in the USA at least. What is also of interest in the Birch data is that the rate of job destruction and creation is higher in those industrial sectors and geographical areas (again the data relates to the USA) where the rate of economic growth is higher. Thus high rates of job loss might be expected in areas where there is a high rate of technological innovation, but this may also be associated with an equally high rate of job creation.

A second important myth is the belief that small firms are more important than large firms in the process of technical innovation. While the role of small firms should not be discounted, recent work at the Science Policy Research Unit at Sussex has shown just how many significant innovations have come out of large companies. At least two things follow from this. One is that policies which are focused primarily on small firms are likely to encourage only a relatively low level of technical innovation. The other is that even if subsidiaries of large companies are attracted to a particular region, and the company decides to develop innovations in that location, there is always the possibility that the firm (likely to be a multi-national corporation) will relocate its plant, or switch manu-facturing to other plants, and the local/regional authority will be powerless to intervene.

Leaving aside the question of the importance of small firms for innovation and job creation, there is also the question as to whether the new technology contains a tendency to reduce or increase the size of firms or establishments. We shall argue later in the book that the information processing capability of new technology may undermine many of the reasons why large firms came into being. We may observe in the future a con-tinued trend in the use of sub-contractors, consultants, and other external agencies, thus reducing the core size of currently large firms. This trend may be accelerated by an increase in telecommuting (working from home via a telecommunications link). However, there are countervailing pressures. The detailed analysis is in Chapters 6 and 7.

New technology, employment, and geography

It has been suggested that microelectronics leads to much greater flexibility in the way in which work is organised, with real choices now being possible. One set of scenarios for the engineering industry suggests that in a country such as Norway, with small communities at the heads of the many fiords, an appropriate use of the technology would be for each community to have some equipment to produce particular parts of a product or to engage in particular parts of a production process. These could be linked up via an information system to form an integrated undertaking. The system would be designed to be particularly user-friendly, emphasising the enhancement of worker skills rather than de-skilling. In contrast there may be certain parts of the world—Siberia has been suggested—where it may be important, because of the presence of energy or raw materials or for other strategic reasons, to have integrated factories where the emphasis is on minimising the number of people employed at that location (Hatvany *et al.* n.d.).

An overview so far

We have put forward a series of arguments about the effects of new technology on employment levels. If there is a theoretical argument justifying the view that microelectronics is a job-killer, it is less obvious and more complex than usually supposed, and by no means universally accepted. In any event, a principal exponent of that theoretical position argues that in practice, in the United States up to the year 2000 at least, the technology will cause net job creation.

Nevertheless, there is bound to be substantial job displacement and creation, and the various tendencies highlighted by the Council for Science and Society (CSS 1981), and by such writers as Mike Cooley (n.d.) and Tony Watts (1983), have to be taken seriously. The wide variety of scenarios should, we argue throughout this book, be treated as a series of options over which we as a society ought to be able to exercise some choice. To do this we need to be alert to the processes going on within society and the economy, what range of alternatives is available, who can and should take part in the decision-making process, and what the effects of the various possible outcomes might be. It is to these issues that the rest of the book is devoted.

3
New Technology and the Experience of Work

Charlie Chaplin's well-known film *Modern Times* portrayed work in modern mass-production factories as devoid of all skill or satisfaction, with workers merely cogs in some vast machine totally outside their control. The new technology of automation had stripped them of any human dignity in their labour. In this chapter we explore a variety of theories and research findings about the relationship between new technology and people's experience at work, concluding that such technology has the potential for improvement as well as for harm. This means that it is important that all those likely to be affected by the introduction of new technology should be involved in the decisions about how it is to be used. Moreover, the impact of new technology on people's experience of work will depend on how they perceive it and the meaning that technology has for them.

Despite the message of *Modern Times* and our images of modern factory work being influenced by our beliefs about what it must be like to work on, for example, the car assembly lines at Detroit or Dagenham, most people, when asked in social science surveys, say that they experience an acceptable level of job satisfaction at work. Surveys over the last two decades, in America at least, have typically reported over 80 per cent of workers being 'satisfied' with their jobs (Hackman and Oldham 1980, 10). To the many people who believe that much modern work is an alienating, tedious, even soul-destroying experience, this is a surprising finding. Is it that different people are turned on by radically different things, and that many people get some enjoyment from performing routine and apparently undemanding jobs? Rather than accepting this rather pessimistic view of human nature, social scientists have suggested two other reasons for surveys reporting such apparently high levels of satisfaction. One is that it is culturally unacceptable to admit that one's job is not satisfying. In a society in which achievement is highly valued, and where people compete for jobs in an apparently open market, people may feel a sense of failure if they admit they have ended up in a job with which they are dissatisfied. Where the boss's response to a complaint might well be 'If you don't like it you can always leave ', then sticking with a job you do not like implies that you lack the motivation to go and find something more satisfactory, or that you are incapable of finding anything better. None of this may be objectively true—alternative, better, jobs simply may not exist even when there is not a high level of unemployment—but to the extent that individual workers believe the dominant achievement ideology in society then they will not wish to 'confess failure' by admitting that they do not enjoy an adequate level of job satisfaction.

A second reason for such high reported levels of satisfaction may be that respondents are answering the question in terms of their level of expectations. Many will have learned that the kind of job that they will probably have to do for most of their working life will be semi-skilled, routine, undemanding, and tedious. Someone with this expectation may therefore compare jobs in terms of level of wages and job security, and not in terms of how intrinsically rewarding the job is. If so, then the response to a question about job satisfaction will be in these terms. They are not dissatisfied with the job content because they did not expect the job to be intrinsically rewarding.

In the light of these difficulties with measuring job satisfaction, academic researchers have made two responses. One is to attempt to devise more sophisticated questions about how workers are experiencing their work, and to give relatively more weight to comparative data: for example, differences in reported levels of satisfaction before and after the introduction of new technology, or between two samples of workers operating different technologies where other factors are constant. The other is to rely on other data which it is hoped might serve as indicators of satisfaction. Levels of labour turnover, absenteeism, and strike activity have been used, but for various reasons none of these measures are at all satisfactory. The assumption is made that dissatisfied workers are more likely to leave the one job for another, take time off, or go on strike, but there has been considerable debate about all three measures, and there are cogent reasons for doubting that any of them serve as adequate indicators of job satisfaction.[1] There are, of course, a number of aspects of work about which a worker might be satisfied or dissatisfied. We have already noted that many workers learn very quickly that the kind of job they are likely to get will not be very satisfying intrinsically: that is, the content of the job will not be particularly attractive. But work involves a variety of factors. It requires the application of effort and, to some degree at least, the exercise of skill and knowledge. It also demands some level of participation in relationships with other workers, perhaps with customers, and, unless the worker is self-employed, with supervisors and management, and at the end of the day work also yields a wage or salary. Each one of these four factors is a potential source of satisfaction and dissatisfaction.

Though people differ markedly, and to some extent systematically, in the relative importance they place on these various sources of satisfaction,[2] our concern here is not with individual differences in sources of satisfaction but with the effect of new technology on the potential sources of satisfaction. Moreover, though individuals may differ in the weight they assign to various aspects of a job, most individuals are prepared to make trade-offs between different sources of satisfaction. An obvious trade-off is that between effort and reward. Assuming that for most people a job which requires a high level of effort (or long hours) is less preferable than one requiring less effort, and that more pay is preferable to less pay, then people will make a trade-off, if they are allowed, between the effort they exert and the pay they are awarded. A reduction in satisfaction on the effort front may be traded for more satisfaction on the pay front. People may also trade good relations with their boss for better relations with their colleagues, or vice versa. We discuss some of the implications of these potential trade-offs in detail in Chapter 6, but it is important to bear them in mind within this chapter as we discuss a variety of theories about the relationship between new technology and job satisfaction.

One further point which needs to be made at this stage

---

[1] For a discussion of the extremely complex relationship between worker satisfaction, labour turnover, and absenteeism, see Ingham (1970), and for a discussion of strike activity rates see, for example, Hyman (1972) or Ingham (1974).

[2] An early, influential, study of differences in orientation to work is by Goldthorpe *et al.* (1968).

concerns the question of ideology, class conflict, and class consciousness. Those social scientists following the Marxist tradition emphasise the twin categories of capital and labour, and take the view that these two major factors of production are in fundamental conflict over the earnings of the enterprise. Any profit accruing to capital is earnings forgone by labour, and vice versa. One highly important way in which capital gains an advantage in this conflict is, they claim, through capital establishing and maintaining a dominant ideology which legitimates the status quo. An important question for those with this Marxist perspective relates to the circumstances in which capital can maintain this ideology and prevent the raising of consciousness of labour as to its own class position. A number of the writers we discuss below have as their primary concern the extent to which technological change will either increase or reduce class consciousness and class conflict. Some of these writers have taken the view that as part of this class war management, on behalf of capital, have been able to steer the direction of technological change along a particular path, stimulating certain developments at the expense of others so that the interests of capital are better served, and the interests of labour are countered.

This approach is elaborated in Chapter 5. Within this chapter our concern is with those, both Marxist and non-Marxist, who have taken as given the particular direction in which technology has developed, and with those who, while accepting that it is possible and desirable to exercise some social control over technology, have taken the view that there is no fundamental conflict at the work-place and that the system of organisation there can be optimised in the light of a set of objectives to which they assume all would be able to agree.

Of the various theories which have been proposed about the relationship between technical change and experience of work, an early and influential, though much criticised, study is that of Blauner (1964). His concern is with the impact of technical change on two of the four factors we previously identified as potential sources of job satisfaction, namely social relations at work and the exercise of skill and knowledge. He accepts the widely held view that the development of mass-production industry produced a sharp deterioration in the conditions of manual work and led to the emergence of a type of worker who combined a generalised hostility to the prevailing society with a primary preoccupation with militant economic action; but he suggests that, as technology continues to develop, the increased use of full automation will eliminate many of the deepest sources of resentment about work, and, in so doing, encourage normative integration within the existing social structure of the enterprise.

He develops his argument by identifying four phases in technological development. Craft production is his first phase; followed by technologies which employ people primarily to mind machines; then mass (assembly-line) production; and finally continuous process production. Examples of these technologies, in their chronological order, would be the pre-Industrial-Revolution spinning and weaving by craft workers; the machine-minders of the early cotton mills; the mass production of consumer goods such as cars and other domestic equipment; and finally the chemical industry.

Among the various characteristics of these technologies which Blauner identifies we highlight three here: the changing skill requirements, the worker's sense of control over the work process, and changes in the meaningfulness of work—these last two characteristics being, to a very large extent, a product of the social relationships which the worker has with management and the end-user of the product being made.

Blauner suggests a U-curve relationship against time for all of these characteristics. For skill, both craft and continuous-flow production require a higher proportion of skilled workers than machine minding and mass production. In the early period of industrialisation it was the skilled (craft) workers who actually carried out the production task, and in the late period the skilled workers constructed and maintained the fully automated continuous-flow production process. In both craft and continuous-flow production the worker has a sense of control over the work process, claims Blauner. In continuous-process plants workers are liberated from the rhythm of the machine and can set their own pace. They are free to move around; they can plan their own work schedules; and they are free to use their own initiative.

Continuous-process technology offers more scope for self-actualisation than machine and assembly-line technologies. The nature of the work encourages a scientific, technical orientation and the changing character of the technology results in opportunities for learning and personal development for a considerable section of the blue-collar force. (Blauner 1964, 174.)

Automation also restores meaningfulness, says Blauner. In the continuous-process plant the worker can get a picture of the complete sequence of operations, and automation reconstructs the collective nature of work, thus encouraging the worker to think in terms of the whole rather than the part.

Blauner also devotes considerable attention to features of technology which imply that workers are likely to feel a greater sense of belonging and identification with factories using highly automated processes. He cites five such features of highly auto-mated plants: the finely graded status structures of such or-ganisations, arising from the more even balance than in other industries between skilled, semi-skilled, and unskilled workers; the relatively smaller size of plants and work units; the changed role of management (with the work units taking over quality supervision and the machinery controlling the level of output, management is left with the job of providing advice rather than instructions); there is blurring of the dividing line between manual and non-manual work; and the greater prosperity and lower proportion of labour costs allows management to offer higher pay levels, a wide range of fringe benefits, and a high level of job security.

A major criticism of Blauner's approach is his assumption that most of those working in highly automated plants have the kinds of jobs he describes: for example, jobs needing a high level of skill, giving considerable autonomy to the worker, and restoring meaningfulness to work.

James Bright (1958), writing at the time Blauner was con-ducting the bulk of his research, reached a radically different conclusion to Blauner, finding that ' . . . automation had re-duced the skill requirements of the operating work force, and occasionally of the entire factory force' (Bright 1958, 8), and a more recent study by Nichols and Armstrong (1976) of a chemi-cal factory in Britain shows just how much routine unskilled labouring work such continuous-process factories still generate.

Nichols and Armstrong claimed that many workers in that factory experienced a high level of alienation from their work.

Bright's study was particularly detailed. He set up a list of twelve components of skill, including such items as 'mental effort', 'dexterity', 'responsibility', and 'decision making', and a set of four broad levels of mechanisation. Though he extended these four levels into seventeen very specific levels of mechanisation, we need not concern ourselves here with that level of detail. His four broad categories he labelled: hand control; mechanical control; variable control, signal response; and variable control, action response. These categories correspond closely to Blauner's, ranging from traditional craft work at the one end to a completely automated, cybernetic, system at the other. The major difference between Bright and Blauner is that, while the latter deals only with total factory production systems, Bright focuses on the individual worker. They agree that skill requirements are reduced as mechanisation moves through the first three stages, but part company in their interpretation of what happens in fully automated plants, Bright claiming that, because the machinery becomes virtually self-sufficient in terms of needing no worker input, such work that does remain is subject to more centralised control and closer supervision even though the tasks to be performed may have become more sophisticated.

A third, early, contribution to our understanding about the possible effects of new technology on people's experience of work comes from research conducted by Joan Woodward in the 1950s and 1960s. Her major finding was that a relationship appeared to exist between the technology an enterprise operated and the organisational structure of that enterprise. This research gave rise to the school of 'contingency theory' which is discussed at some length in Chapter 6. Of particular significance to our discussion in this chapter is her conclusion about the way in which technology appears to influence organisational structure at the two ends of the scale of technological complexity. She suggests that, whereas in the simplest forms of production the organisation was necessarily subservient to the needs of technical co-ordination, in process production technical co-ordination is incorporated in the machinery itself, and the organisation is subservient to the social needs of the people working in it. ' . . . in the technically advanced firm organization serves primarily social ends, its function being to define roles and relationships within a social system. This means that the organisation planner can concentrate on establishing the network of relationships which is best for people'—Woodward (1965, 123), quoted, as is much of the above summary, from Gallie (1978, 14). The chief executive of the continuous-process plant, claimed Woodward, is no longer preoccupied by technical matters, but issues of organisation, industrial relations, and public relations. Moreover, process production produces a clearly defined primary task for both management and men—that of keeping the plant running as near full capacity as possible.

Another important element in Woodward's explanation, and one that we will be returning to in Chapter 6, is the impact of technology on the control system within the factory. She defines control as the task of monitoring the outcome of activities and if necessary taking corrective action (Woodward 1970, 38). Two main dimensions of control systems are distinguished: the extent to which they are unified or fragmented, and the degree of impersonality of the system. Her theory is that the more that control is personalised, the more likely it is that there will be friction between management and the workforce, but that continuous-process plants, because they are so highly automated, have unified impersonal control systems which minimise conflict and lead to more satisfying working conditions.

Serge Mallet (1969), writing in the context of developments in political theory in France, and hence particularly concerned with issues of class consciousness and conflict, took a view directly contrary to Woodward, arguing that new technology was leading to the development of new forms of class conflict that would threaten fundamentally the stability of the existing structure of the capitalist enterprise. Many aspects of his analysis were similar to those advanced by Blauner, but they led him to an opposite conclusion. Workers, he claimed, would be *objectively* integrated within the enterprise for three reasons: a change in the character of the salary system, the nature of the worker's skill, and the high level of job security.

With automation, output from the plant does not depend directly on the effort of any individual worker, Mallet suggests.

Therefore individual work measurement becomes inappropriate and each worker's salary is determined primarily by the overall economic situation of the enterprise. As Gallie puts it, summarising Mallet, 'Integration, in the sense of identification with the enterprise and its overall performance, becomes then a necessary corollary of the desire to further one's own interests' (Gallie 1978, 18).

The change in the nature of the workers' skill comes about, claims Mallet, because old craft-based skills of general applicability are replaced by skills needed to operate specific pieces of machinery. Again quoting Gallie's summary of Mallet:

The new emphasis is on attentiveness, and the ability to take responsibility for machinery that is both extremely expensive and highly dangerous. It is a skill that can only be learned through experiences with a specific complex of machinery and, in contrast to that of traditional skilled work, it is specific to a particular enterprise. (Gallie 1978, 18.)

Having developed such plant-specific skills, a worker will not be in a position to find a job of equivalent pay elsewhere, and so this factor, too, 'integrates' the worker into the plant, in the sense Mallet uses the term.

Mallet's third factor, job security, follows from his first two points. With the company having invested heavily in plant-specific training for a particular worker, and output from the plant not being directly related to the number of workers in the plant, firms operating these technologies will be reluctant to hire and fire such highly trained workers in line with fluctuations in product demand. Such reluctance is reinforced by the relatively small savings which would be made by reducing labour costs, as the capital intensity of production is so high, and also by the fact that there is less opportunity to shed labour in the downturn when using highly automated plant because running such plant at all usually requires a certain fixed minimum number of workers. 'The workforce thus comes to view employment in the firm as a permanent career, and it sees its future as intimately tied up with the fate of the enterprise' (Gallie 1978, 18).

In sharp contrast to Blauner, who argues that this 'objective integration' will lead to social integration of workers and management, Mallet argues that this objective integration has the opposite effect, leading to a new form of revolutionary consciousness that aims at the overthrow of the existing pattern of social relations in the enterprise. A major reason for this is that the high productivity of fully automated plants satisfies the basic consumption needs of workers but, instead of this leading to a high level of satisfaction, workers freed from the immediate concern of making ends meet are then able to pose what Mallet takes to be the more fundamental problem of their alienation from their work. Their ability not only to begin to understand the nature of their alienation but to learn to do something about it is also enhanced by the development of automation. The job security which Mallet, as we have already indicated, assumes will follow the introduction of fully automated plant gives workers the opportunity to develop a substantial knowledge about the firm and its activities. Secondly, responsibility in such plants becomes decentralised as a consequence of the firm's dependence on the worker's use of his initiative. This gives the workforce both a much greater capacity to coerce management and a consciousness of its collective power, claims Mallet. Hence workers in plants using the most advanced technology have both the awareness and the ability to mount revolutionary demands, and, moreover, suggests Mallet, their activities will act as a model and inspiration for the wider working class.

A second French critique of the notion that new technology is likely to enhance people's experience of work is the work of Pierre Naville. Where Mallet began in agreement with Blauner but came to different conclusions, Naville starts from a quite different position, arguing that work itself, under automated technology, is inherently unsatisfying. It deprives the worker of any contact with the raw material, destroys any residual sense of a personalised relationship with the machine, and is frequently boring, lonely, and mentally stressful, he asserts.

In support of these assertions he compares the situation 'in traditional industry', where, for the worker arriving in the factory in the morning, 'his individual machine is a bit like his child. He goes to work, and when he arrives he gets it working; when he leaves, at the end of the day, it sleeps until the next morning. It is his machine.' The situation is altogether different in a continuous-process industry. There 'when . . . a worker arrives at work and replaces a colleague while the production line continues to operate, he feels that this machine is not really his. He arrives, he checks it over; then he goes away, and the machine continues to go on running without him.' (Gallie 1978, 21-2, quoting Naville 1963, 203-4.)

Naville also identified a series of problems with work with automated technology which he characterises as arising from a marked increase in mobility. This comes about, he suggests, for three reasons: there is frequent movement of personnel between different posts; the supervision and direction of a complex set of machinery encourages the multi-functional use of in-dividuals; and there is an increase in the use of shift work. This increased mobility is seen by Naville as a source of great dissatisfaction. Not only does automation lead to a heightened sense of alienation, it also, Naville suggests, enables workers to obtain a better understanding of the real determinants of the pattern of authority relations prevalent in the society, which Naville presumes to be dictated by social class rather than technocratic or efficiency considerations. This is for two reasons. Firstly, Naville argues, like Joan Woodward, that automation allows a high degree of autonomy to the system of social or-ganisation of the plant. For Naville, given his Marxist theor-etical perspective, this visible autonomy sweeps aside the veil and reveals the nature of the existing patterns of hierarchical organisation in their true light—as patterns that are socially, and not technologically, determined.

Secondly, argues Naville, there is a growing contradiction between the requirements of the work process and the existing system of social relations:

Efficiency can no longer be achieved through the division of labour, but through the development of more efficient patterns of co-opera-tion. This requires a closer integration of the social organization of work, an integration not only at the level of relations between workers, but also between the different hierarchical levels. The traditional patterns of social organisation with their emphasis on hierarchical differentiation then become a major obstacle to the most efficient use of the technology. (Gallie 1978, 24.)

Naville is more cautious than Mallet in the conclusions he draws from his analysis about the extent of conflict with management likely to be brought about as a result of extensive use of such automated technology. The negotiating position taken up by the unions will be an important factor, and Naville recognises that the behaviour of the unions will be influenced by many factors additional to the technology used in some trade union members' firms. He also suggests that

their demand for control cannot be restricted to one of self-government within the enterprise, but it will have to aim at the development of a complex institutional system that will be able to strike a balance between the need for worker autonomy within the enterprise on the one hand, and the need for economic co-ordination at higher economic levels on the other. (Gallie 1978, 25.)

Gallie summarised the views of Blauner, Mallet, and Naville because he wished to set up an empirical test of their theses. During 1971 and 1972 he surveyed four oil refineries—two each in France and Britain—interviewing a total of over 800 workers. His conclusion was that his data 'provided support for neither Mallet nor Blauner' (Gallie 1978, 298). Their view that the main traditional sources of grievance in industry—salaries, and problems deriving from the nature of the work task and of work organisation—would cease to be of importance as a focus of grievance in continuous-process industries turned out to be unsupported. He found that the cultural and institutional differences between France and Britain had a more important influence on people's experience of work than the common technology each factory used. Thus, with regard to level of pay, 'the British were broadly speaking content; the French dissatisfied. This was not simply a marginal grievance for the French workers: it was the dominant source of their resentment about their situation as manual workers in French society.' (Gallie 1978, 296.) Moreover, whereas the British workers were also overwhelmingly satisfied with their payment structure, the French were deeply convinced that their payment structure was based on unfair principles.

Similarly with regard to attitudes to management, the British workers fitted the Blauner model, expressing a high level of contentment with existing procedures of decision making and a consensus about organisational goals, whereas the French saw their enterprises as socially dichotomous and exploitative.

Both Gallie, and Nichols and Beynon (1977), agree that

the description of work in a continous-process plant given by Blauner and Mallet is misleading. The technology does not create the almost Utopian working conditions they describe, and people's work tasks and conditions are not so very different in continuous-process industries as compared to earlier technologies. If there are to be significant changes in people's experience of work in the future, these are, concludes Gallie, much more likely to be brought about by changes in cultural values, in institutional arrangements, or in management attitudes or trade union objectives (Gallie 1978, 317).

It may be that, in his desire to show how little support his findings give for the Blauner and Mallet theses about the determining influence of technology, Gallie underplays the role technology can play. The design of his study, comparing plants in different cultures and countries rather than comparing plants operating different technologies, is bound to heighten the reader's awareness of cultural rather than technological influences. Moreover, though he devotes several pages to a description of the work done in these oil refineries, Gallie does not address the question of whether the technology itself differs in the French as compared to the British plants, nor whether details of the work organisation itself differ in a non-random way between the two countries. Clearly technology is not the only determinant of work organisation, or of people's level of satisfaction at work. Cultural and institutional factors play an extremely important role. It is, however, unlikely that such a central part of the production process as the production technology itself has no influence at all on how people experience work. We therefore need to explore further, and at a more detailed level, what relation can be identified, within a given cultural and institutional milieu, between technology, work organisation, and people's experience of work. Such knowledge would be useful not just to enable us to forecast when the revolution might happen, but also in helping us to know how to make choices between different versions of a particular technology, and how to organise work to operate that technology. We turn now to those who have had this specific concern.

The socio-technical systems perspective

Immediately following the nationalisation of the coal-mines in Britain at the end of the Second World War a team of social scientists from the Tavistock Institute in London began an extensive series of studies of work organisation in that industry. The significance of their research for our study of new technology at work is twofold. Firstly, the results of their work are of enormous interest in themselves. Secondly, this work diffused in a number of different geographic and theoretical directions. It was the foundation of what became known as 'socio-technical systems theory' and had a direct influence on the Scandinavian experiments in work organisation, of which the SAAB and Volvo cases described in Chapter 5 are just two examples. The Tavistock Institute, and in particular the lead researcher in the coal-mining studies, Eric Trist, were also founder-members of the so-called Quality of Working Life movement, the impact of which has been particularly great in North America.

The Tavistock study was of new ways of organising work as novel coal-getting technology was introduced into pits in the 1950s. Although the technology was not microprocessor-based, and the events are now thirty years past, the study's theoretical approach and empirical findings are of such interest and relevance that it is worth describing them in some detail.

Their point of departure was the pre-mechanisation tradition of 'single place' working in the Durham and Northumberland coalfields. Under this system of working a face was mined by a 'marra group' of between two and six men. ('Marra' is Geordie dialect for mate. Readers of D. H. Lawrence will be more familiar with the word 'butty' which has the same meaning.) Each man was a 'complete miner', carrying out all aspects of face work—the preparation work, the cutting of the coal itself, and the advancing of the face ready for the next cut. If one person performed all these tasks then they were defined by the Tavistock group as having a 'composite work role'.

Some faces would be worked by one man alone, others by two together—though each would have a 'composite work role'. In other words, there would be no division of labour. Some mines operated a two-shift system, others had three shifts. All those across the shifts who worked that face were members of the marra group. They shared the same paynote and were self-selected. Each group had the opportunity of changing its membership at quarterly 'cavils'—events jointly supervised by management and union when faces were allocated to gangs by lot in order to even out differences in the ease of winning coal from particular seams. The relationship between each group and management was akin to that of contractor and client. There was a price agreement for a specific task, either coal-getting or doing the necessary stonework in preparation, and Trist *et al.* (1963) observe that management dealt with workers through the union to a far greater extent than in most industries. As a result there was only a vestigial junior management system. The first-level supervision, known as the deputy, stood in a service rather than an executive relation, his responsibilities being for safety, maintenance of supplies, and shot-firing rather than for direct supervision. The next level up in the management hierarchy, the overman, did not often visit a man in his work-place.

Trist *et al.*'s explanation for this form of work organisation focuses on aspects of the task of coal-mining. The production machinery has to be moved every day as the face moves, geological variations mean that the size and quality of the seam is unpredictable, and there is always the possibility of the collapse of the walls, roof, etc. This 'threat of instability from the environment makes the production task much more liable to disorganization' (Trist *et al.* 1963, 20). Also, because mobility in the pit is so difficult, supervisors can visit the operators only occasionally throughout the shift. Hence there is much less control by management than in a factory yet far more events about which decisions have to be taken. Therefore, argue Trist *et al.*, the work group has to be given responsibility for the entire cycle of operations, and for handling the interdependence between those on different shifts. The group has to be self-regulating. Trist, in an earlier publication with Bamforth (1951), coined the term 'responsible autonomy' for this degree of self-regulation.

The introduction of face conveyors (an early stage of mechanisation in the pits) in the early part of the twentieth century led to the adoption of longwall cutting. Rather than one or two men working at an isolated face, up to fifty men would be involved in winning coal by working at a face up to seventy to ninety metres in width. The form of work organisation adopted by management in the Durham coalfield, at first, was to formally separate the three major production processes of preparation, coal-getting, and advancing (the props, conveyors, etc.) and to set up distinct groups, each working a whole shift, to perform each process. The cyclical character of the work now acquires central importance; under this system a right balance is needed between manpower and task size, and there is a need to schedule operations so each process is completed in each shift. Not only did management set up functionally differentiated work groups but they also split up jobs within each group into such constituent elements as cutting, scuffling, drilling, hewing, filling, pulling, and stonework. It is not important to understand what each of these jobs entailed. The point is that such an elaborate division of labour was set up despite, as Trist *et al.* observed, each job not being so 'complex that the average faceworker could not, in a reasonably short time, be expected to become qualified in more than one' (1963, 47). Moreover, the research team observed that the skills required for these jobs were much less than the skills needed to contend with interferences emanating from the underground situation. In other words, there are many tacit skills of a general nature which faceworkers develop. But the conventional longwall work organisation fails to build these tacit skills into the work groups.

With such an elaborate division of labour, the conventional longwall system requires much more integration than single-place systems but, claimed Trist and his colleagues, there were no mechanisms available to provide this. It could not be done by direct supervision because of the inability of management to be present at the face frequently enough. Also, each manager only had shift responsibility and so could not co-ordinate the work across shifts. It was not possible to expect the group to act with 'responsible autonomy'—to be self-regulating—because of the elaborate division of labour. The third mechanism, that of the price agreement, upon which management placed much reliance in single-place working, also broke down because of the detailed breakdown of jobs. A great deal of time was spent haggling over prices for specific tasks. Also, with no common denominator for fixing rates, different criteria were used for measuring performance—tonnage, yardage, cubic measures, number of operations completed, etc.—and so contradictory interests developed.

In response to these difficulties a different form of work organisation, deriving from the single-place tradition, was developed. Known as composite longwall working, it was based on a self-selected group of forty-one men, who allocated themselves to tasks and shifts and received a comprehensive payment on a common note (Trist *et al.* 1963, 77). This wage was split into two parts: 58 per cent was a basic wage to cover everything, so bargaining with the deputy was virtually eliminated, and the remaining 42 per cent was a bonus on output. This form of work organisation, an attempt to reintroduce responsibility, autonomy, and the client–contractor element to the management–worker relationship, had the following four aspects:

(a) Composite work method—oncoming men take up the cycle at the point left by the previous shift.

(b) Composite workmen—each person is multi-skilled. Though all members of a cycle group do not need to be competent in all five of the recognised face skills, the team as a whole must be able to deploy sufficient resources on each shift to man the roles likely to arise.

(c) Composite work groups—self-selection and internal allocation of roles.

(d) Composite payment—common paynote in which all members share equally, comprising shift-rate covering all work done, plus bonus based on coal produced. Provision is also made for payment when the bonus cannot be earned because of circumstances outside the workers' control. As always with social science experiments, it proved exceedingly difficult to measure the effects of these changes. The researchers reported a more positive attitude to work among those operating the composite longwall system, with a higher level of what they termed 'mental health'— fewer symptoms of stress and less conflict. They also reported a substantial increase in productivity.

There was an increase of 20 per cent in the number of cycles completed per unit time after reorganisation from the conventional to the composite longwall system (a cycle being the three processes of preparation, getting, and advancing), though they do note that, at the same time, better conveyors were introduced and on one face there was a change from hewing the coal to cutting it (Trist *et al.* 1963, 256).

The researchers go on to draw some general conclusions from the results of this study. One conclusion is that they note that further mechanisation (beyond the composite longwall described above) was having major effects on work organisation. The higher level of automation involved fewer, less narrowly specialised, work roles and fewer task groups, each undertaking a large part of the cycle. Therefore the problem of integration is lessened. This results in a less complex work organisation with more in common with single-place and composite systems. This is a conclusion to which we return, and challenge, in Chapter 6.

A second conclusion is that despite the findings from small-group research it *is* possible for relatively large primary work groups of up to fifty members to operate successfully. The coal-mining studies provided evidence that groups of this size could be maintained and could sustain self-regulation. It should be added, though the authors do not mention this, that the highly solidaristic ideology of mining communities in the North-East of Britain may well be of great importance in sustaining these egalitarian working groups and that they may not be viable in other areas where a more individualistic ideology is adhered to.

Their third conclusion is that 'when new technology is introduced an inappropriate form of work organization—that associated with the technical trial of the machinery—tends to be carried over to subsequent operational units' (Trist *et al.* 1963, 293). Therefore there needs to be a period to allow exploration of satisfactory forms of internal work organisation when new technology is introduced, and this period may well need to be longer than that usually allowed for new skills acquisition.

The result of this research was the establishment by the Tavistock Institute of socio-technical systems theory. As Trist defines it:

The concept of a socio-technical system arose from the consideration that any production system requires both a technological organization—equipment and process layout—and a work organization relating to each other those who carry out the necessary tasks. The technological demands place limits on the type of work organization possible, but a work organization has social and psychological properties of its own that are independent of technology. A socio-technical system must also satisfy the financial conditions of the industry of which it is a part. (Trist *et al.* 1963, 6.)

This perspective has had a remarkable impact on the development of our understanding of work organisation and job design, and is at the root of much of experiments and innovations in work organisation in Scandinavia. It does, however, suffer from at least two weaknesses. One is the lack of awareness that the technological organisation is itself subject to social forces. The other is that no consideration is given to conflicting interests and the possibility that different forms of work organisation may be preferred by one group in the enterprise rather than another. We explore both these points in the next chapter.

The quality of working life movement

Now an evangelistic organisation which includes governments, managers, and trade unions, the Quality of Working Life (QWL) movement has as one of its founding fathers Eric Trist, the principal investigator of the coal-mining study reported above. Its most comprehensive statement of aims, theory, and practical guide-lines is given by J. Richard Hackman and Geoff R. Oldham in their book *Work Redesign* (Hackman and Oldham 1980). In it they state their belief in motivation through the design of work and set out a model for good job design which is reproduced in Figure 3.1.

The model is developed from research they have conducted into the relationship between job characteristics, motivation, and job satisfaction. They are particularly concerned with factors which lead to individuals wanting to do jobs well because of the intrinsic satisfaction from doing the work. This they term 'internal work motivation' to distinguish it from motivation which arises from extrinsic factors such as a payment-by-results wage system. Earlier work by Turner and Lawrence tested the theory that if jobs were carefully designed to incorporate

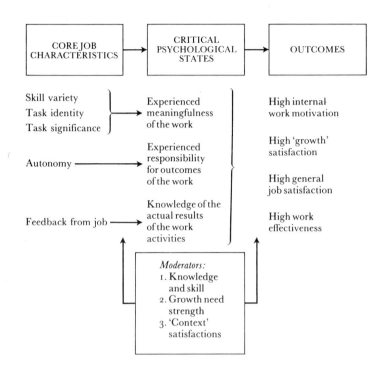

Fig. 3.1 The Job Characteristics Model.
(from Hackman & Oldham, *Work Redesign*, 1980, Addison–Wesley, Reading, Massachusetts; p.90, Fig. 4.6. Reprinted with permission.)

specific features then high levels of satisfaction and motivation would result. They predicted that the amount of variety in work, the level of employee autonomy in performing the work, the amount of interaction required in carrying out task activities, the level of knowledge and skill required, and the amount of responsibility entrusted to the job-holder would be positively associated with satisfaction and a low level of absenteeism. They found, however, that this relationship only held up for workers in small-town settings. Workers in urban areas displayed no relationship between these job factors and absenteeism and, perhaps more surprisingly, those jobs scoring high on variety, autonomy, responsibility, and so on were those providing least job satisfaction.

The Hackman and Oldham model therefore introduced two important refinements to the model. What matters, they claim, is the psychological state of the worker and the skills he or she possesses. High internal work motivation and satisfaction result from experiencing meaningfulness and responsibility from the work, and from obtaining knowledge about the actual results of the work activities. This experience and knowledge can come, as the model in Figure 3.1 shows, from characteristics of the job, but only if certain 'moderators' are present. If these moderators are present (we discuss what these are in a moment) then work will be experienced as meaningful if it requires a variety of skills from the worker, if a complete piece of work is done (task identity), and if the worker sees the task to be of significance. Jobs which are designed to give the worker autonomy will produce the experience of responsibility, and designing in mechanisms to provide feedback from the job provides knowledge of the actual results of the work activities.

However, well-designed jobs will not give workers experience of meaningfulness or responsibility if the so-called moderating factors are not present. If workers lack an appropriate level of knowledge and skill to complete the set task then they will experience frustration rather than meaningfulness in the face of demanding jobs. A second important moderating factor is the strength of workers' need to 'grow' in the job. Workers who have adapted to tedious and undemanding jobs and therefore have no desire to enhance their skills and knowledge are, it is suggested, likely to resist rather than welcome jobs with the suggested core characteristics. Thirdly, if 'context' satisfactions are low (for example, there is poor pay or threats of redundancy) then performing a job which is well designed will not be enough to generate high motivation or intrinsic job satisfaction.

Using this model Hackman and Oldham claim to have had some success in redesigning jobs to give the outcomes they predict. The implications of their model for the relationship between new technology and people's experience of work are of course clear. They would suggest that, when new technologies are being introduced, care should be taken to design jobs associated with those technologies which have the core job characteristics they specify. Hackman and Oldham are, however, pessimistic about the amount of attention that will be paid to job design considerations as new technology is introduced. They suggest a number of reasons for this. Firstly, they admit that, although there is now a substantial body of knowledge about what makes a job a good one, there is much less knowledge available so far about how to apply job design theories in real life settings. Secondly, social scientists are only just beginning to develop procedures for evaluating the economic costs and benefits of innovative work design; and, thirdly, little is known about the conditions under which these innovative job designs persist across time and diffuse across companies and countries.

Moreover, say Hackman and Oldham, even if we had this knowledge, it would still be a struggle to implement good job design principles. It will always be a battle between 'hard' engineering knowledge about the specific new technology being implemented and 'soft' behavioural knowledge. The styles and values of both employees and managers currently support a job design philosophy that has its roots in work study and industrial engineering practice, and it flies in the face of conventional wisdom to believe that jobs can actually be designed rather than simply be the result of technologically determined requirements.

Supporting Hackman and Oldham's pessimism are the results of a large-scale experiment in the United States, along the lines of the Tavistock study reported earlier in this chapter. This, too, was in the coal-mining industry. The programme began in 1973 as an attempt to incorporate the most recent findings about job design into the work organisation, and with the intention of improving the quality of work life. By 1978, however, the programme had disappeared. The social scientist charged with monitoring and evaluating the project (Goodman 1979) refused to answer the question as to whether or not the experiment had been a success, asserting that this was the wrong question. What should be learned from the study is why certain things worked and others failed, so that other quality of working life programmes might do better.

Hackman and Oldham suggest that to reverse current trends

and move towards good job design practice requires that organised labour should make QWL a high-priority item, that managers discover that QWL pays 'in coins they value', that government should decide that it requires, or is to encourage, organisations to improve QWL, or that the cultural climate must change.

In contrast to Mallet and Naville, they do not appear to believe that there is any fundamental conflict of interests between capital and labour which would lead to them expressing opposing preferences with regard to the taking up of the job design principles they advocate. As we shall see in the next chapter, there are those who believe that management have often adopted forms of work organisation which give rise to unsatisfying jobs because it is cheaper for them so to do. Hackman and Oldham admit that no piece of research has yet shown a positive relationship between high job satisfaction and high productivity. One explanation for this may be that for many jobs it is high productivity which leads to low job satisfaction. If management can obtain high levels of output from individual workers by designing jobs, such as those on assembly lines, which force people to work hard then output may be high at the expense of job satisfaction. An element of job satisfaction for many may be the absence of coercive pressure from management and the ability to carry out one's job at one's own pace, a pace which may be slower than that which management would prefer. It may therefore be that, if a technique were ever found to evaluate the economic costs and benefits of innovative work design, it might reveal that in a number of cases such innovation had net costs, to management, rather than net benefits. The question then would be how unions assessed the costs and benefits and whether they would want to pressure management to bear this cost in order that their members could enjoy the benefit of higher job satisfaction—possibly at the expense of their receiving lower wages.

4
New Technology, Work Organisation, and De-skilling: The Theory

Introduction

We now turn to an alternative, fairly recent, intellectual tradition which emphasises the conflict of interests in the workplace and suggests that management, on behalf of capitalist interests, have moulded technical developments so that they suit these interests and run counter to the interests of labour. Technological developments have been used down the ages to fragment jobs and de-skill workers, they claim. This has enabled management to get more effort out of labour and pay a lower wage. Because management use the technology to exercise tight control over workers, managers have not needed to pay much attention to issues of job satisfaction and motivation because only when workers have autonomy does their level of motivation become a relevant consideration. This form of analysis has been applied to various periods during the industrial revolution and to more modern developments.

Early industrialisation

Perhaps the most dramatic shift in the pattern of work associated with new technology was that which characterised the first industrial revolution, as thousands moved to the newly forming industrial towns to work in the mills. For most of the two centuries since that phenomenon began the conventional explanation has been that it was the new technology (the spinning-jenny, the new weaving frames, and the steam engine) which gave rise to the factory. More recently this argument has been stood on its head with the suggestion that it was the rise of factory production which caused the technological developments which have given us today's machinery. Marx was wrong, it is now claimed, when he asserted that the steam engine gave us the capitalist. It was the capitalist who gave us the steam engine.

This argument was first put forward by a Harvard economist,

Steven Marglin, in an article he provocatively entitled 'What do Bosses do?'. It is a re-analysis of the reasons for the industrial revolution in Britain.

The cotton industry in Lancashire, in North-West England, was the seat of the industrial revolution and the birthplace of large-scale factory production. Here were the archetypes of William Blake's 'dark satanic mills'. Before the rise of factory production, spinning and weaving were done on a putting-out basis, with individuals or groups of spinners and weavers working in their own cottages in country towns, villages, and hamlets dotted throughout rural Lancashire. Merchants would travel around on a regular basis giving out raw materials and collecting the spun, or woven, product. As the demand for cotton goods rose (partly owing to economic growth caused by advances in agricultural techniques and extended foreign trade) cotton production shifted from the putting-out system to mills being set up in the rapidly urbanising towns. The reason for this shift, according to Marglin, was not the higher efficiency of factory production compared to home-working under the putting-out system, but because the merchants could not force the home-workers to work hard enough to meet the increased demand for cotton. The merchants had a problem, claims Marglin, because the spinners and weavers had what the economists term a 'leisure preference'. In other words, they preferred, when they had earned a certain basic sum of money sufficient to meet their immediate needs, to spend the rest of their time in leisure rather than in work to increase their earnings. As they were paid on a piece-work basis, this meant that there was a ceiling to the amount of output any one person would do. Once a sufficient weight of cotton had been spun or woven to generate the needed income, work would stop. If the piece-rate was raised in an attempt to persuade the workers to produce more output, exactly the opposite effect was obtained as individuals would only produce the (lower) quantity required to meet their cash target. If piece-rates were lowered in an attempt to boost output then presumably workers would take other jobs, or would switch to another merchant.

Faced with this failure to raise output through the wage rate, it became advantageous to the manufacturer to set up central spinning and weaving factories in order to reduce the rising cost of merchants travelling ever more extensively in search of additional putting-out workers. It also—and perhaps more importantly—gave much tighter control over workers, enabling the factory owner to offer an all-or-nothing contract to the mill-workers, such that they would be required to work the then standard fourteen-hour day, six-day week—a far more intensive work-rate than that chosen by the putting-out workers who could exercise their own preference in the trade-off between work and leisure.

It is in this context that many of the early technical innovations in the textile industry developed. The millwrights and other inventors who were making these technical developments were not operating in a vacuum. They were making and altering machines for specific customers, and the people who were prepared to pay for improved, labour-saving, machines at this time were the owners of the factories. Hence, it is argued, the technology was developed to meet the requirements of large-scale factory production rather than of small-scale craft producers operating out of their own cottages. The large-scale factory, it should be emphasised, had not emerged because there were economies of scale in spinning and weaving. The factory was not necessarily more efficient than the home-based weaver. The factory had come into being, Marglin argues, because it enabled the owner to force higher levels of output out of operators. But, once there, the factory created the demand for technologies which were appropriate for factory production, and these technologies almost certainly enhanced efficiency. As individual craftsmen neither had the desire nor the capital to pay for development work on machinery suitable for small-scale home-based use, factory-based manufacture would have become relatively more efficient over time as the technology was developed. Evidence in support of this view is quoted by Charles Sabel (1982, 39):

The evidence is that the discovery of basic principles of machine design, their application to large-or small-scale industrial production, and the creation of efficient managerial techniques were all independent of the creation of the factory system by nineteenth-century capitalists. The most basic principles of machine design, for example, were discovered by Renaissance and Baroque instrument makers, military engineers, and scientists. Their application to such industries as spinning and weaving in seventeenth and eighteenth centuries often resulted in inventions suited to the circumstances of petty producers: The new machines required little capital and a family-sized labor force, and hence were well suited to the perpetuation of cottage industry.

It is clear from this that it was social and not technical factors which were responsible for the emergence of factory production.

Marglin argues that the organisation of work within factories was influenced by a further set of social factors—the need for the factory owners to maintain a monopoly over the knowledge needed to run the enterprise. In any large-scale undertaking there is usually some advantage in subdividing the work so that individuals can develop special skills in particular tasks. Marglin suggests that factory owners subdivided tasks way beyond what was necessary to maximise the development of individual skills and minimise time wasted by individual workers having to perform more than one operation. They subdivided the work further in order to ensure that there were aspects of the job that only the owners knew how to do. They could thus prevent any group of workers going off and setting up their own workshop in competition to the merchant and capitalist class.

He concludes, therefore, that for the twin reasons of the capitalist's desire to force high output from individual labourers and to prevent labourers from setting up in competition, a pattern of work organisation emerged in the early cotton industry that was characterised by intensive working in factories, with individuals performing fragmented tasks. To suit this pattern of working, technology suitable for use in factories rather than workshops, and for use by unskilled workers rather than craftsmen, was developed. Given the substantial investment in developing that technology over many years, this form of working is now probably the most efficient way of running this industry. It does not, however, mean that this is the only way in which the industry could have developed. The technology is not a neutral determining force, but is itself subject to human choice which is made by the powerful on the basis of what suits their interests. How persuasive is this argument in explaining the development, use, and effects of technology now, in the microelectronics era?

An extremely influential protagonist of the view that these arguments still hold was Harry Braverman, whose book, *Labor and Monopoly Capital: The Degradation of Work in the Twentieth Century* (1974), gave birth to a sunrise industry of scholars engaging in development and critique of his argument. Not an academic himself, Braverman began his working life as an apprentice coppersmith and spent much of the rest of it in socialist journalism and in publishing. His arguments about the use of new technologies in the factory and the office are therefore based on a lifetime of work experience.

Braverman's thesis is that there has been a general and progressive de-skilling of jobs in the twentieth century, and that there is a long-term trend for jobs to be stripped of intrinsic content, and become more routine. He argues that under capitalism the interests of workers and capitalists are fundamentally opposed so that the latter, in order to pursue their objective of profit maximisation, need to exercise tight control over the former. One technique of central importance for doing this is Taylorism—the method of so-called 'scientific management' invented by Frederick Winslow Taylor at the turn of the century. The essentials of Taylorism involve the use of work-study engineers to study in fine detail the work an individual is doing; through detailed analysis develop the most efficient method of doing that work and establish a 'fair' time for performing particular elements of the task; develop a specialised staff function whose task is to allocate jobs to individual workers; and monitor performance against the rated time for the job. Elements of this technique have been widely adopted in manufacturing industry, and work study, time-and-motion analysis, and payments-by-results schemes are all children of this 'scientific management' approach.

Taylor himself argued that his methods operated in everyone's best interests. No worker, he suggested, was clever enough to be able both to perform their own job and to analyse the best way of doing it. It was therefore of benefit to any individual worker to have their job analysed by an expert and be shown the most efficient way to do it. This would enable them to produce more, without a commensurate increase in effort, and thus earn a higher wage or salary. Managers would have more predictable work flow, and owners would earn higher profits as their share from the increased efficiency of operation.

Braverman takes a contrary view. He holds a rather higher view of individual intelligence than Taylor, arguing that the person who has to do that job every day is the one most likely to know how to do it. Braverman's view is that work-study engineers are not contributing knowledge by their analysis of others' work: they are acting as eyes and ears of management to find out on their behalf how quickly a job can be done. Just as the capitalist desires maximum output from labour for least cost, so does the labourer want maximum pay for the least effort. Given this opposition of interests, management will desire better knowledge of worker performance in order to ensure higher output. The fundamental task of the work-study engineer is to provide, in this battle of wits between management and workers, that knowledge for management so that they can exercise tighter control. Braverman goes further and suggests, with Marglin, that management will also endeavour to develop methods of working that involve operators in fragmented, or short cycle time, tasks for reasons other than efficiency. One reason for this is that it is easier to measure, and therefore control, performance if management can observe precisely what workers are doing. Another is that unskilled or semi-skilled operators can be recruited to perform these fragmented tasks, replacing the skilled craft workers who had previously performed the unfragmented tasks. Leaving aside the fact, to which we will return in a moment, that less skilled people can usually be paid less wages than those who are skilled, an important characteristic of craft workers is that they often exercise tight control as an occupation over the job that they do. Skilled craft workers frequently establish their own methods and pace of work, and this control by the occupational group is usually backed up by the craft union. Semi-skilled or unskilled workers usually find it harder to find a basis upon which to exercise occupational control and are more often either non-union or members of a general union which does not afford a platform to develop such control.

With regard to pay, the benefit to management of de-skilling so that the elements of a job are fragmented into their constituent elements is that those elemental jobs can be assigned to those with skills, and hence pay rates, to match them. Braverman (1974, 81) quotes J. R. Commons' description of a meat-packing conveyor in the USA in the early part of this century to illustrate this point:

It would be difficult to find another industry where division of labor has been so ingeniously and microscopically worked out. The animal has been surveyed and laid off like a map; and the men have been classified in over thirty specialties and twenty rates of pay, from 16 cents to 50 cents an hour. The 50-cent man is restricted to using the knife on the most delicate parts of the hide (floorman) or to using the ax in splitting the backbone (splitter); and wherever a less-skilled man can be slipped in at 18 cents, $18\frac{1}{2}$ cents, 20 cents, 21 cents, $22\frac{1}{2}$ cents, 24 cents, 25 cents, and so on, a place is made for him, and an occupation mapped out. In working on the hide alone there are nine positions, at eight different rates of pay. A 20-cent man pulls off the tail, a $22\frac{1}{2}$-cent man pounds off another part where good leather is not found, and the knife of the 40-cent man cuts a different texture and has a different 'feel' from that of the 50-cent man.

Another central element in Braverman's thesis is that management, in order to achieve their objective of tight control over labour, not only fragment jobs horizontally—in the sense that different physical elements of the same task are split off from each other to be done by separate individuals—but fragment jobs vertically, as well. This vertical fragmentation splits the conception of the job from its execution. Whereas a traditional craftsman would decide what should be done, how it should be done, when it should be done, and to what level of quality, as well as actually executing the task itself, modern management within capitalism takes on all these former conceptual functions, leaving labour merely to carry out the mechanical aspects of performing the manual task. Braverman quotes the editorial of a 1918 edition of the *International Molders Journal*, noting that 'A half-century of commentary on scientific management has not succeeded in producing a better formulation of the matter' (Braverman 1974, 136):

the gathering up of all this scattered craft knowledge, systematising it and concentrating it in the hands of the employer and then doling it out again only in the form of minute instructions, giving to each worker only the knowledge needed for the performance of a particular relatively minute task. This process, it is evident, separates skill and knowledge even in their narrow relationship. When it is completed, the worker is no longer a craftsman in any sense, but is an animated tool of the management.

The implications of Braverman's views about the de-skilling and fragmentation of work for the development of new technology parallel Marglin's: they would both agree that the pattern of technological development is profoundly influenced by managerial requirements. While Marglin emphasises the way in which technology has been developed to meet the requirements of large-scale factory production, Braverman suggests that technological developments are also shaped by management's desire to split conception from execution. The conceptual, thinking, aspects of the job are automated and incorporated into the machine, leaving the worker as a relatively mindless machine operator—a human, though merely mechanical, appendage to the machine. He illustrates this with an historical account of the development of the computer controlled machine tool.

Conventional machine tools, such as lathes and milling machines, are frequently operated by highly skilled craft workers. Working just from engineering drawings, the skilled machinist will decide the sequence of cuts to be made, how fast he will run the machine, and how much material to take off on each cut (the 'speed and feed'), and will control manually each of the levers and screws on the machine which control each aspect of its operation. An early exercise in 'scientific management' by Frederick Taylor was to investigate optimum speeds and feeds for various metals, machine tools, and cutting tips. These were recorded on charts and he recommended the appointment of speed and feed men who would ensure that the optimum rates were adhered to. This was an early example of the erosion of the skilled machinist's job. The more major fragmentation occurred as developments were made in the control of the machine tools themselves. In the 1950s a control system based on punched tape was developed. This enabled someone to establish the numerical co-ordinates of the movement of the cutting tool over the entire sequence of operations and to transfer this to code on punched tape which was then fed into the machine. The machine would read the punched tape and automatically move the tool, control the speed of rotation of the chuck of the lathe, or of the milling tool, and carry out the work while the operator stood by checking that no malfunctions occurred. Such machines were described as 'numerically controlled' or NC machine tools. The advent of relatively cheap microelectronics meant that by the late 1960s it had become possible to control these machines by dedicated mini- or micro-computers directly attached to them, such machines having 'computer numerically control' or CNC. The vexed question has always been: Who should write the programs which control these machines? Braverman, and many others following in his tradition since, have argued that machine tools should continue to be operated by skilled craftsmen after the introduction of CNC, and that the programs should be written by those who operate the machine. In practice, in many instances companies have fragmented the machining tasks as CNC has been introduced, setting up separate programming departments, often staffed by people from a 'white-collar' background. CNC machine tool manufacturers have often promoted their products with the claim that they can be operated by unskilled operators, and it was widely believed for some time that engineering companies were taking the opportunity given by the introduction of CNC machine tools to replace skilled craft workers by unskilled or semi-skilled operators on the shop-floor, and putting white-collared technicians in the programming office. The advantages to management of so doing are claimed to be that wage costs are reduced, that the relatively powerful skilled craft union is replaced by the weaker general union, and, perhaps most importantly, management can exercise much tighter control over what happens on the shop-floor. The tighter control is thought to come about because many of the decisions about how work should be done are now taken in the office rather than on the shop-floor. An individual machinist might be tempted to choose feeds and speeds that were either faster or slower than those appropriate for the job in hand. This might be to speed the job up to earn a larger bonus or to create some leisure time during the working shift, or to slow the job down so that an impression of busy-ness is created and he is not given further work. There may also be short-cuts that the machinist could take in doing the work which may mean slight deviations from the specification in the drawing. All these possibilities are reduced if the program is prepared off the shop-floor by those with no direct interest in shortening or lengthening the time taken to do the job for personal reasons.

David Noble (1977) takes this argument a stage further and argues not just that the use, but also the design, of the CNC machine tool has been influenced by these management control considerations. In the early days of the development of control systems for machine tools, two rather different principles for control were being explored. The first was the one we have just described, in which the path of the cutting tool is analysed by a technician from inspection of the engineering drawing, described in formal mathematical language, and recorded on punched tape. The other system is known as record–playback and involves the skilled craft operator going through the physical motions of machining the first piece of the batch with the machine in record mode. The control system of the machine tool records all the movements of the cutting tool and then, when put in playback mode, reproduces the actions previously performed by the skilled operator. The latter type of control system thus retains the skill of the craftsman, does not interpose a technician into the work process, and thus does not fragment and de-skill the task. Noble argues that the technical performance of the two systems in their development phase was similar but that the NC system was adopted in preference to the record–playback system because managements preferred the tighter control given to them by the NC system (Noble 1977).

An obvious criticism of the Braverman de-skilling thesis comes from the data showing a growing percentage of the workforce with formal qualifications. Braverman himself presents figures (Braverman 1974, 379) showing the massive increase in the relative number of white-collar workers in the workforce. In the United States they have moved, according to his figures, from 15 per cent of the workforce in 1900 to 35 per cent in 1970. Even the number of craft workers has risen from 10.5 per cent to 13.9 per cent over the same period (1974, 427). To these figures Braverman has two responses. The one is that the use of white-collar staff rather than blue-collar workers does not imply that the former are doing 'complete' jobs over which they have full control. We have already seen that white-collar staff were introduced into the engineering industry as programming adjuncts to CNC machine tools. Their tasks are, in Braverman's terms, fragments of the previously integrated craft job of machining. Many other white-collar jobs, claims Braverman, are similarly fragmented. He describes as an example the large drafting and design rooms of engineering companies as having

been organized, in many instances, on the same principles as the factory or office production line, and . . . staffed by serried ranks of detail workers whose pay scales, if they are better than those of factory operatives or clerical workers, are perhaps not so good as those of craftsmen, and who dispose of little more working independence and authority than the production worker. (1974, 406.)

His second response to the apparent increase in the number of skilled and semi-skilled workers this century is to challenge the concept of skill implied in these data. To some extent, the apparent general increase in skill levels of the working population is the result of bureaucratic reclassification of occupations—in the 1930s the US Bureau of Census decided that all machine workers should be termed operatives and deemed semi-skilled. In addition, being termed 'skilled' often simply means that the worker is carrying out a job traditionally performed by someone who has undertaken an apprenticeship and is therefore a craft worker by training. It is not necessarily an accurate description of the demands of the job, and the skill label may result more from the strength of the trade union or occupational group in maintaining that its members should continue to hold that position, than from a dispassionate assessment of the skill content of the job. Jobs may thus, in many instances, have been de-skilled to the extent that management has tighter control through fragmenting the tasks and more tightly specifying what should be done, though not to the extent of eliminating craft workers from the job.

Implications for organisations

If there is an intrinsic conflict between capital and labour leading management to seek out means of exercising tight control over workers through fragmenting and de-skilling jobs, then this will have implications for the way in which whole enterprises are organised. This in turn is likely to affect the development and use of new technology, not just at the work-place but throughout the enterprise as a whole.

Braverman's radical analysis of work organisation at the operational level has been extended by a number of writers concerned to explore its implications for these wider or-ganisational issues. They challenge two assumptions commonly made, either implicitly or explicitly, in conventional or-ganisation theory: (a) that an institution adopts that form of organisation which best fits its technology, environment, and other contingent factors, the 'best fit' being in terms of that organisation structure most suited to co-ordinating the tasks performed by members of that institution; (b) that members of the institution have entered freely into a contract with that institution to provide a service to that institution in return for reward, and that they accept as legitimate both the institutional goals and the means of co-ordination used to achieve those goals (rational–legal authority).

The specific implications of these analyses for organisation theory have been developed by writers such as Benson (1977a, b), Goldman and Van Houten (1977), Heydebrand (1977), Clegg and Dunkerley (1980), and Clegg (1981), though in a rather general way.

The overall conclusion to these studies is that organisations under capitalism are likely to have a more elaborate division of labour, hence a more extensive organisational hierarchy, make greater use of rules and procedures, and generally have a more complex form of organisation than would be the case if there were not a fundamental conflict being worked out within the firm between capital and labour. The argument is as follows, and emerges from the Braverman thesis.

Firstly, for the Tayloristic reasons discussed earlier in our exposition of Braverman, management is likely to fragment and de-skill tasks, splitting off conception from execution. This implies a more elaborate division of labour: horizontally, in terms of there being larger numbers of people each doing dif-ferent jobs at the work-place; and also vertically, as new jobs are created among white-collar workers to co-ordinate and control the various semi-skilled or unskilled workers—a task made necessary because of the deliberate fragmentation of that labour. The operation of 'scientific management' through Taylorism itself creates a more complex organisational structure. The operations of this scientific management structure itself need to be managed, and so a further management hierarchy has to be erected on top of the workshop-based organisation structure to manage the 'scientific managers'.

Secondly, it has been suggested that a second organisational control device used by management in the face of potentially recalcitrant workers is the use of the internal labour market. Kathleen Stone (1973) identifies this device in operation as early as the late nineteenth century, in the steel industry. Up until that time much of the work in steel-making had been done by individuals or groups of steel-making master craftsmen on a jobbing or sub-contracting basis. The owners of steelworks would contract with a master craftsman for a particular job to be done at a particular price. That master craftsman would then himself employ helpers to assist him in the completion of that contract. The position of the master craftsmen was maintained through a strong steelworkers' union. The owners were not content for this much power and control to reside in the hands of the craftsmen, and Stone details the strategies they used to break the union and replace the jobbing/ sub-contracting arrangement by an employment relationship. Among these was the introduction of job ladders. The work was fragmented and the various new jobs which were created in turn demanded different skill levels and responsibilities to be exercised. The innovation was that management arranged these various jobs in such a way that an individual might expect to progress from one to another if their performance merited it. Such progressions were termed 'job ladders', and economists describe the way in which people are selected and promoted up these ladders as internal labour markets. The difference between the fragmentation described by Braverman and that analysed by Stone is that the former came about in order to de-skill and control through Tayloristic practices, while the latter came about partly in order to create a range of jobs which could be formed into a job ladder. This is an additional reason for management to create an even more elaborate division of labour than might be warranted from mere technical considerations of the tasks to be done. Both reasons for fragmenting work imply the creation of a managerial hierarchy to co-ordinate and control the various fragmented jobs. Where job ladders are created, further managerial work is involved in managing the operation of the internal labour market. Thirdly, it is suggested that other organisational devices, which we discuss in more detail in Chapter 6, are also used by management in order to extend their control over labour. For example, it has for a long time been generally accepted by students of organisation that any organisation is likely to need a number of rules and procedures to guide the behaviour of organisational members. There has been much debate about how much such rules and procedures should be used: too much, and the organisation becomes immobilised in bureaucratic red tape; too little, and the organisation descends into anarchic chaos. The conventional wisdom has been that the nature of the job being done by the organisation determines the requisite amount of rules and procedures. If tasks are usually routine and predictable, then management can co-ordinate matters most efficiently by setting up rules and procedures to instruct people what to do. If the tasks to be done are more complex, and surrounded by a higher degree of uncertainty, then co-ordination may be done more efficiently through instructions and information passed down the managerial hierarchy. There are implications from this analysis for the application of information technology within organisations which will be explored in Chapter 6. The point of introducing these ideas here is to draw attention to the radical analysis which suggests that management may use rules and procedures, and other more complex organisational arrangements such as product groups or even matrix organisational structures, not just because they may be a more efficient means of co-ordination but because they are necessary if capital is to control recalcitrant labour. They may, for example, be an adjunct to the use of Taylorism, where fragmented work leads to the use of particular types of organisation structure, whereas co-ordination of the same overall task, if it were carried out by craft workers on an integrated basis, might be carried out by means of a much simpler organisation structure.

Fourthly, the problem is not just a mechanical one of co-ordinating and controlling the extra flows of information required when jobs are fragmented. If organisations operated through the exercise of what Alan Fox has termed 'high trust' relationships between both managers and operators, and among operators themselves, then much of the necessary co-ordination and flow of information could be handled informally and on the basis of good will (Fox 1974). It is the argument of Braverman and some other radicals (though not of most of Braverman's critics, as we shall see in the next chapter) that within capitalism the inherently antagonistic relationship between capital and labour inevitably generates a 'low trust' relationship. The inability to operate with high levels of trust itself implies the use of more complex formal organisational arrangements as a replacement for the more informal co-ordination mechanisms which are possible when there is a higher level of good will.

We conclude, therefore, that the radical perspective on the labour process offers a far-reaching critique of conventional organisation theory. As we intend to use this theory in Chapter 7 to explore the implications of new technology for the shape of organisations, it is important to develop our understanding of organisational processes in the light of this radical attack. We attempt to do this in the next chapter, but before embarking on this task we must note some of the criticisms which have been made about the Braverman thesis and its implications.

The response to Braverman

Though the majority of those reviewing Braverman's work took a positive view of his thesis, there have been a significant number of detailed criticisms and revisions to his argument. In the remainder of this chapter we deal with five of those that have been put forward by those who give general assent to his radical perspective. These are: that Braverman ignores worker resistance; that management may be ignorant of the most effective ways of meeting worker recalcitrance; that there are more ways of killing a cat than skinning it (in other words, that there are mechanisms other than fragmentation and de-skilling which management can use to control recalcitrant labour);

that technological and market opportunities vary across firms and that these influence the outcome of the struggle between capital and labour over the form of work organisation which is adopted; and that the tight control of labour and the need to maximise output from individual workers may not be management's most dominant concern in every case. Additional criticisms, some from outside the radical perspective, have been made, but we deal with these in the next chapter where we put forward an alternative conceptual framework within which to consider the issues. In the meantime, let us look at the above five sets of objections in more detail.

As Stephen Wood points out in his Introduction to his edited book discussing the de-skilling thesis (Wood 1982, 15), Braverman appears to assume 'that prior to Taylorism control was not in the hands of management, but was rather a kind of management by neglect (the *laissez-faire* method, as Taylor called it), in which workers knew more than managers'. In fact we now know that in many instances, for example in the steel-making industry that Kathleen Stone describes, control was firmly in the hands of the craftsmen themselves. We should thus expect resistance from workers to attempts by management to wrest control from them, and the historical evidence for this is now forthcoming. An implication of this is that the resulting forms of work organisation, and hence the type of technology developed, reflects the compromises management have had to negotiate with workers. Edwards (1979, 16), for example, characterises the situation as one where for most of the twentieth century management have met 'chronic resistance to their effort to compel production' and consequently 'have attempted to organize production in such a way as to minimise opportunities for resistance and even to alter workers' perceptions of the desirability of opposition'. In the face of such opposition, those managements with the wit to explore alternative strategies have adopted a range of techniques. In his study of labour relations in the motor industry in Coventry in the 1960s, Andy Friedman identified two dominant management control strategies. In small firms doing sub-contract work for the large car assembly plants, where the level of unionisation was weak, and management hired and fired labour in line with demand as it rose and fell, the form of work organisation approximated more closely to that which Braverman describes. Friedman termed this the 'direct control' strategy. Within the large car plants themselves, where the unions were strong and the companies were able to offer a relatively stable employment policy by shifting the troughs and peaks in the work-load out to sub-contractors, management had adopted what Friedman termed the 'responsible autonomy' strategy. This strategy

attempts to harness the adaptability of labour power by giving workers leeway and encouraging them to adapt to changing situations in a manner beneficial to the firm. To do this top managers give workers status, authority and responsibility. Top managers try to win their loyalty, and co-opt their organizations to the firm's ideals . . . ideologically. (Friedman 1977, 78.)

Richard Edwards draws a more complex picture. He argues that management has developed a range of responses and has attempted to maintain control through the use of industrial relations procedures, through bureaucratic rules which channel conflict into manageable and acceptable ways, and especially through the use of internal labour markets and dual labour markets which divide and segment the working class. Even Edwards, according to Wood, still presents management 'as omniscient, conspiratorial and able, at least for a certain period of time, to get its own way—that is, to solve successfully its problem of control' (Wood 1982, 16). A more satisfactory perspective, Wood argues, is to view the resulting production system as the product of a struggle between contending groups, a struggle which, moreover, is conditioned by the particular product and labour markets, technology, and trade unions which are surrounding and are a part of the conflict.

Craig Littler and Graeme Salaman (1983) go further. They suggest that in many circumstances, and particularly more recently, the central problems facing management are not so much to do with control over labour but are much more to do with such matters as obtaining orders for products, getting the design right, innovating, and handling their relations with the capital market. Profits for the individual firm, they argue, often depend far more on these factors than on the surplus value to be exploited from their own workers. They will thus not bother to pursue de-skilling strategies with great vigour as the return on vigorous managerial activity is greater in other areas of operation.

The conclusion we reach, which we share with Wood, is that 'the quest for general trends, such as progressive deskilling of the work force, or general conclusions about the impact of new technologies are likely to be both theoretically and practically in vain' and that 'to incorporate worker resistance, labour and product markets and extra-economic factors involves more than simply extending one's analysis; it amounts to a theoretical reconsideration' (Wood 1982, 18, 22). In the next chapter we review a number of recent empirical studies of the introduction of new technology and associated changes in work organisation, and these substantiate our conclusion that there is no general trend, either toward or away from de-skilling. The introduction of new technology creates the opportunity for a wide range of choices to be made about how work should be organised. The remainder of this book is devoted to putting forward further theoretical considerations and to suggesting ways in which analysis can be carried forward about some of these opportunities and effects of new technology at work. What these effects eventually turn out to be will, of course, depend on the strategies which the various parties choose to pursue, the prevailing capital, product, and labour market circumstances, and the technological choices which the scientists and engineers make available.

5

New Technology, Work Organisation, and De-skilling: The Evidence

Having reviewed the theory and some descriptions of the impact of technical change in previous decades and centuries, we now turn to the results of current research into the impact of microelectronics-based technology on the work that people are doing today. Most of this chapter is devoted to a description of five sets of case studies of the new technology, showing the wide variety of different impacts it can make on the nature of work.

Robots in West Germany

The first of the case studies we describe is of the use of robots in a West German factory and gives strong support to the de-skilling hypothesis described in the previous chapter. The researcher conducting this case (Wobbe-Ohlenburg 1982) comments that much of the publicity about using robots in factories emphasises the way in which they can make life more pleasant for humans by taking over the dirty, dangerous, or monotonously repetitive jobs. Viewed in this light, robots might well be seen favourably by workers and their trade unions, and thus welcomed onto the shop floor. However, when the researchers reached the shop floor they discovered that robots were far too expensive for management to waste them on unskilled jobs. Humans were still doing these jobs: the robots had taken over the skilled elements of the task.

A typical case was that of an arc welding operation. The complete task involved lifting a number of panels from their storage racks, loading them on to a jig, clamping them into position, arc welding a seam to join them, and then transferring the welded sub-assembly from the jig to another storage rack so that it could be transported to the next production stage. Rather than having the unskilled parts of the operation—the lifting and placing—performed by the robot, and the skilled part—the welding—performed by the welder, the opposite solution was adopted. The robot did the arc welding, serviced by a human who lifted and carried for 'him'. In the pre-robotised days the job had a number of characteristics which job design experts would have recognised as leading to a reasonably high level of job satisfaction. It involved a mixture of tasks and required a reasonable level of skill. It was paced by the worker, and the person occupying that job belonged to a skilled occupational group strongly represented by a craft union. After robotisation the job became less skilled, was machine paced in that the cycle time was dictated by the speed at which the robot did the welding, and the human part of the job was filled by a semi-skilled operator belonging to a general trades union. This union had to represent a wide range of occupations and was therefore less able to protect that particular worker's specific set of interests. This particular type of robot application was not untypical, the researchers found.

The researchers concluded that the decision to use robots in this way was because management was much more concerned with cost and product quality than with the quality of the working life of the welder. Not only were robots, at the time of the study, uneconomic to use on routine unskilled work, but they were more reliable than humans in performing skilled tasks. Therefore, not only were the more skilled elements of the tasks robotised first to save the wages of the more highly paid skilled labour, but by using robots management could be more certain that a critical weld on a critical component was being done to specification. Robots, although they develop faults, do not usually miss out welds, or weld inaccurately, on a random basis, and it is possible to have them wired up so that when they do malfunction supervision can be immediately informed. Moreover, it is also possible that management took the view that unskilled workers could be controlled more easily than skilled workers because they had less powerful union backing, and that they could be replaced more easily because there was a bigger pool of labour to draw from. Eliminating skilled labour would thus give more power to management.

It is not surprising, perhaps, that management gave prime concern to the two factors of cost and product quality over that of work-life quality. The important point to emphasise here is that, because of management's concerns with cost and quality, the application of robotics technology did not lead, as much publicity would suggest, to an improvement in work-life quality. In fact it led to the opposite. What this case also illustrates is that there are different possibilities in using the new technology and there may be a conflict of interests between workers and management in the way it is used. There may therefore be bargaining about how the technology is used, and the outcome will depend to some extent on the bargaining strengths and skills of workers and their union representatives.

Biscuit making in Scotland

The second case study is reported by two Scottish researchers, David Boddy and David Buchanan of the University of Glasgow (Buchanan and Boddy 1983). This provides less support for a simple de-skilling hypothesis, revealing more of the complexity and variety of outcomes associated with new technology. Within a book detailing several fascinating cases of the implementation of new technology and its organisational impact, they detail two applications of computerised equipment controls to a biscuit making line in a United Biscuits factory in the UK. One application was to the biscuit dough mixing process, the other to the weighing of packed biscuits. Of particular interest is that these two applications, within the same factory and installed by the same management team, had radically different impacts on the experience of work of the operators. This was not because of any difference in management intentions. Indeed, there seems to have been no thought given at the design stage to the impact of the technology on people's work experience. The sharp difference in impact on work experience between the two applications was an entirely fortuitous outcome.

The application with the particularly detrimental effect on operators' jobs was the biscuit dough mixing automation. This automation took place in two stages, the first occurring in 1971 and the second, the focus of Buchanan and Boddy's study, ten years later. It is a little difficult to disentangle from Buchanan and Boddy's description the particular effects at each stage of the automation, but the differences between the pre-1971 and post-1981 situation are clear. Prior to 1971 the mixing was carried out under manual control by time-served master bakers controlling a team of operators. Mixing was done in open vats like large domestic food mixers. The doughmen could tell by the feel of the dough, and by the sound that it made during mixing, if it was too dry or too wet, and add small amounts of required ingredients to compensate. If equipment failed then the nature of the fault was usually clear to the master bakers. After 1981 the operation was controlled automatically and monitored by two semi-skilled mixer operators. The computer stored biscuit recipes on punched paper tapes which, when fed into the computer, put the correct amounts of ingredients into the appropriate mixing machines. The computer controlled not only the mix recipe but also the time, speed, temperature, and timing of all the mixing operations and indicated when the operator should add the various 'sundry' ingredients (those that had to be added in small quantities). All mixing now took place in fully enclosed machines, and faults in the process were flashed onto a 'mimic board' by the computer.

The automation thus de-skilled the job both in terms of how the job was classified and in terms of how much knowledge was required to do it. Semi-skilled operators were now used instead of master bakers and the computer took much of the knowledge element out of the job. Also, the way in which the machinery was designed (by totally enclosing the mixing vessels) meant that operators lost much useful feedback about the state of the process.

Buchanan and Boddy conclude that:

The job of the mixer operator thus had the following characteristics:
1 the operators had no need of traditional baking skills and had an incomplete understanding of the production process and equipment;
2 the rest of the production process was dependent upon this stage, but the operators got only intermittent feedback on the performance of the mixing process and were unable to effect control over it;
3 the operators appeared to have difficulty visualizing the consequences of their action or inaction on later production stages;
4 the operators could not trace and diagnose equipment faults, skilled technicians were necessary, and repairs took longer;
5 it was repetitive and boring and the operators became apathetic and careless;
6 the operators did not develop knowledge or skills that would make them promotable. (Buchanan and Boddy 1983, 189.)

Emphasising the judgement that the job was considerably worse after automation are quotes from both management and operators. From management:

There was more variation in the things he did. Now they seem to lose all interest. There are three or four operators who've been here longer than the computer, and I can see they've really switched off. I've seen it happen to new people coming in as well. It destroys the human contact. (1983, 187.)

And from an operator:

It's much less interesting, more routine, very little scope for human error now. Initially, it's more skilled, till you get to know the set up, then there's a fair amount of boredom. Except when something goes wrong. (1983, 189.)

The other new technology application was to the weighing of packed biscuits. The weight of packets of biscuits coming off the line was a matter of some importance to the company. European Economic Community legislation allowed variation around the nominal weight (that printed on each packet) so long as the average for a particular production period equalled the nominal weight. There was thus a legal obligation to ensure the packets did not fall below the required weight, but on the other hand the company did not want to give free biscuits to customers by making the packets heavier than necessary. Prior to the installation of the new technology the biscuits, after being wrapped, went on to a 'checkweighing' device. The biscuit packets travelled on a conveyor over this machine, and if a packet did not fall within pre-set weight limits it was pushed off the conveyor, later to be opened and good biscuits fed back into the line. Because the checkweighing machine captured no information on packet weights someone from quality control took packets from the line at random and calculated their average weight. If weights were seriously adrift, plant management were informed so that corrective action could be taken. Such corrective action could take place at two points in the process. At the baking stage the ovensman could adjust the temperatures in the zones of the oven to correct thickness, colour, and the moisture content of the biscuit, the first and last of these factors affecting the weight of the packet of biscuits. Adjustments could also be made at the wrapping stage. The wrapping machine was pre-set to wrap a given number of biscuits. If the biscuits coming through were each rather heavy then the machine could be adjusted to wrap fewer biscuits per pack. With the conventional weighing system it took some time for the information about packet weights to get to the point where decisions could be made to adjust either the oven or the wrapping machine.

The technological change made was to take a commercially available checkweigher and replace the standard electronics with a microprocessor controller. Packets leaving the wrapping machine still passed over the checkweigher as before but now, instead of merely passing or rejecting each pack, the computer captured the data it measured and performed various analyses including the average weight and standard deviation for each hour. This information could then be used by those controlling the production process to take appropriate corrective action. The wrapper could be quickly adjusted to wrap more, or fewer, biscuits, and the ovensman had a video display unit (VDU) next to his control panel which gave a summary of the information from the checkweigher, updated every two minutes. This enabled him to adjust his controls immediately the information had been captured. As Buchanan and Boddy conclude:

The [microprocessor-based checkweighing] package only captured, processed and distributed information about packet weights. It gave operators and management timely information that they used to alter the running characteristics of the line and make it more cost effective. The package itself did not control the line, nor did it tell operators and management what action to take when things went wrong. Operators and management still had to use their understanding of the process to control it. (1983, 181.)

The ovensman felt that the new package had increased the interest and challenge in the job. For the weighing machine operators the new technology had increased communication with the ovensman, and led to smoother production and less pressure from management. Though there were some criticisms of the new technology, the general consensus was that it had improved the quality of the jobs people did.

Manufacturing in the Midlands

The third set of case studies we discuss here was carried out by Barry Wilkinson of the Technology Policy Unit at the University of Aston in Birmingham. Four firms were studied: an electroplating plant; a rubber-extruding factory; a firm making optical lenses for spectacles; and a machine-tool maker. Wilkinson takes issue with the view that new technology makes an *impact* on the work that people do. He insists, rightly, that the way in which technology is used is the result of a number of factors, not least of which is the political process between management and workers. The results he reports shows very wide variation in the way work was organised after new technology had been introduced on the shop-floor. At one extreme was an electroplating plant where management took the view that operators were unintelligent and unreliable. When the process was automated, management took care to place the controls on the other side of a brick wall from the process machinery, in a locked outhouse. The keys for this were held by only three people—the works manager, the general manager, and the chief maintenance engineer. At the other extreme, a machine tool company was considering the purchase of a (CNC) machine tool which would have the facility for manual data input (MDI). With MDI the skilled craftsman operating this machine would be able to write his own program to control the machine and enter it directly into the machine's computer. Both these examples illustrate particular choices made by management about the organisation of work after the introduction of new technology. Wilkinson's third case, a rubber-moulding company, exemplifies his contention about the political nature of the process of establishing new patterns of work organisation. In this factory, management had bought within the space of two years two new, more automated, sets of machinery. In both cases the machinery was introduced straight from the supplier to the shop-floor. This meant that all the necessary development and adjustments for full-scale production had to be carried out with the full involvement of the production workers. Wilkinson records that:

In the manufacturing manager's own words: 'Development in the production environment meant that the operator would productionise

machinery, and rather than management giving production para-
meters to the operators, the operators would to some extent give them
to management. (1983, 50.)

Advocates of participative approaches to the introduction of
new technology would applaud the involvement of production
workers in 'productionising' the machinery, arguing that such
involvement would mean that the particular knowledge of the
workers, gained from their on-the-job experience, could be
made good use of, and also that such involvement usually leads
to greater commitment to the change. However, in this case
management saw such involvement as a disadvantage. They
would have preferred process control and development staff to
have established the new processes, and would have preferred
to recruit 'green labour' to the new machines so that 'bad
habits' would not have been carried over from the old production process. Management claimed that it was 'difficult to
convince the men that the new processes were far more 'delicate', and mistakes far more costly. Bad habits on the old presses
could be tolerated, but with the new ones 'the whole process
has to be just right'. (Wilkinson 1983, 112.)

However, the major difficulty for management was that of
re-establishing piece-work rates on the new machinery. Because
the operators had been so heavily involved in the 'productionising' of the new technology and were thus extremely
familiar with the technical possibilities of the machinery, they
were in a strong position to negotiate higher rates than the
time-study person would normally allow.

For both these reasons, management in this company decided that in future productionisation would be carried out
away from the shop-floor. They built a separate development
area, and Wilkinson reports that at the time he was writing his
book three major new projects were being developed within it,
but as none had reached completion there is no data about the
effects of this new pattern of introducing new technology in
that company.

Wilkinson's fourth case was that of an optical company,
making spectacle lenses. In this company, although management had an enlightened attitude to issues of work organisation, the pattern of work resulting from the introduction
of new technology was particularly de-skilling. Management's

practice was to update machinery continually and to buy 'the best and the most modern' (Wilkinson 1983, 43). Decisions about the design of such machinery were made by the suppliers; there seemed little variation in the machines on offer in their implications for de-skilling; and the result was a gradual process of de-skilling of most of the production workers. Management appeared to regret this, to some extent at least because they themselves were master craftsmen and committed to the notion of craft skill. The production manager, for example, complained that, ' . . . there's no training needed to operate this, so it's de-skilling the job and I don't like that, I like to train somebody . . . ' (Wilkinson 1983, 45). Moreover, management did not take the opportunities presented by the new machinery to tighten control over production. They could, for example, have easily concentrated skills into the hands of a very small elite of technicians, and transformed the operators into unskilled, cheaper, and replaceable labour, but they did not. As compared with the way in which computing is usually taken out of the hands of machinists when CNC machine tools are introduced, consideration was given in the optical company to the idea of persuading the operators who cut the surfaces of the lenses to do some of their own computing, and even to the idea of installing computing facilities in the surfacing room itself. Also, the speed of the production process could have been more tightly controlled. The machines could have been used to pace the workers, or management could have monitored the utilisation rate of the various machines. They did not even consider doing this.

Instead, management's response to the de-skilling effects of the new technology was to introduce a system of job rotation. Within each section of the company workers were now rotated from job to job automatically, regularly, and compulsorily, and this system covered a range of levels, not just the most de-skilled of the jobs. The system was introduced by the general manager who had arrived in the company only three or four years previously, bringing with him a strong philosophy of participation and involvement. Wilkinson points out that not only did the system of job rotation create immediate psychological advantages (' . . . the staff like it, they like the change . . .' was one comment), but there were also direct economic advantages for management. Absences due to sickness and holidays could easily be covered because most people now knew most jobs and could easily be switched around. Also, such a multi-skilled workforce meant, claimed Wilkinson, that a shiftwork system could be used with a minimum of supervision. The firm was considering a two-shift system and the general manager considered that the introduction of job rotation was vital preparation for this.

One of Wilkinson's conclusions from his study was that there is enormous variation in the forms of work organisation possible with new technology. Though firms go to considerable lengths in making calculations to justify the efficiency gains in introducing new technology, their *post hoc* evaluations are usually extremely vague. They therefore cannot know the relative efficiency of different forms of work organisation, and these may in fact not have very much influence on levels of efficiency. Therefore there is considerable room for choice in the form of work organisation to be adopted, and the one that suits management's often unexamined assumptions about the best way does not have to be chosen. Wilkinson illustrates his conclusion by suggesting that a comparison of the optical company and the plating company displays the folly of the notion that de-skilling and increased management control over production processes necessarily leads to increases in efficiency. He argues that in the case of the optical company the return of control to the workers ensured a more effective productive unit. While his conclusions may be correct—and we discuss this at greater length in the next chapter—his evidence, unfortunately, does not go far enough to give weight to them. We do not have any data about the performance or the market position of the two firms. If, as is possible, the optical company was in a monopolistic or oligopolistic position, they could afford to adopt work practices which were actually rather costly even though they gave higher levels of satisfaction to both workers and management. This does not mean that better forms of work organisation should not be introduced simply because they are more expensive. That would be akin to justifying slavery on the grounds of economic efficiency. What is in question is the comfortable argument that socially acceptable forms of work organisation always just happen to be those that are the most efficient. It is more likely that in many situations there is a conflict of interest between capital and labour over this issue and thus a need to negotiate the outcome.

CNC machine tool use in Britain and West Germany

The first case we reported was from West Germany, and the next two from the UK. We turn now to a series of cross-national comparative studies carried out with great care and detail by researchers from the Henley Management College in Britain, the International Institute of Management in Berlin, and the Laboratory for the Sociology of Work, Aix-en-Provence, comparing forms of work organisation in British, West German, and French factories. In the case of Britain and West Germany the study was extended to include the relationship between new technology and work organisation. The Anglo-French-German study was not principally about new technology, but its findings are of great relevance to the subsequent Anglo-German study specifically concerned with work organisation and CNC technology in manufacturing industry.

The former study compared carefully matched samples of British, French, and German factories, and having held as constant as possible factors such as size of enterprise, technology, and the firm's environment, discovered substantial differences in the way work was organised in the three countries, concluding that these could best be accounted for by features of each nation's culture, and in particular its educational and occupational structures. Caricaturing their findings somewhat, they discovered that West German firms had flatter, more widely based pyramids of management hierarchy than French firms, and that British firms tended to have forms of work organisation that were not pyramidal because there was extensive use of line and staff relationships. They attributed these differences primarily to the fact that in West Germany there was an extensive system of vocational training, further in-service training, awarding of qualifications, and positioning in the hierarchy on the basis of such qualifications. This meant that management were able to push decision making well down the hierarchy, giving those actually operating the CNC machines much autonomy, and expecting a high level of integration of the various functions such as quality control, work planning, programming of the machine, and machine setting to take place via those operatives actually on the shop floor. In France, by contrast, there was a polarisation of skills, characterised by a high level of expertise among the managerial cadres and technicians, and a very low proportion of skilled craftspeople among the shop-floor workers. This led, they argue, to a situation where decisions were made by the elite at the top of the management hierarchy and then fed down the line of command. Workers under this arrangement required much communication with management and close supervision, and hence rather narrow spans of control and a long managerial hierarchy.

The British case represents the kind of compromise commentators on our national characteristics have come to expect. British factories contained an elaborate mix of skills and occupations. There was in the UK only a moderate amount of vocational training and relatively few people with formal technical qualifications. In particular, there are few highly qualified technical managers. The result was that there was not the level of ability on the shop floor to sustain the German system of decentralisation and shop-floor level integration of work. Neither was there the level of ability among an elite of managers for decision making to take place at that level in the organisation and then to be passed down by functionaries, as in the French system. What happened was that the various decisions were taken by technicians and others in staff positions and fed into the line management structure. We thus had what Sorge *et al.* termed a polarised and differentiated system (1983, 162) with the shop-floor and direct production functions having to relate to management supervision, maintenance, preparatory and planning functions, and technical design and development, who each gave instructions and advice from their staff positions.

The Anglo-German study of microelectronics in manufacturing (Sorge *et al.* 1983) was a study of the use of CNC machine tools in the engineering industries of the two countries, and it showed the extreme flexibility of CNC technology. 'There is', Sorge *et al.* concluded, 'no effect of CNC use as such' (1983, 147). Instead, they suggested five different factors influencing the relationship between CNC use and work organisation. These were:

1. Company or plant size:
2. Batch size, or time needed to machine a batch:
3. Type of cutting and machinery:
4. National institutions and habits of technical work, management, and training:
5. Socio-economic conditions of the present situation, regarding shortage of natural resources, limitation of mass markets, and slow growth. (Sorge *et al.* 1983, 148.)

The relationship they found between work organisation and plant size was in line with that predicted by organisational contingency theory, namely that the small plants had very unbureaucratic, simple forms of organisation. There was no systematic apprenticeship training in the plants, but a lot of pragmatic and flexible learning in how best to use CNC. Larger companies had more systematic procedures and a greater division of labour. This meant that programming was done away from the workshop, in specialised departments. It was not the case, though, that large firms had the most sophisticated technology, nor that high technology itself signified greater bureaucratisation. The reverse was found to be the case in some instances, with some larger companies being slowed down in their introduction of new technology because of bureaucratic organisations and industrial relations arrangements.

Processing large batches meant that firms tended to split programming from machine operating, but not necessarily to create separate departments for programming. The explanation given for this splitting of conception from execution with large batch sizes was that 'the smaller the batch size is, the greater is the need for frequent conversion of machines to new tools, fixtures and parts programmes, and the less is machine-setting expertise differentiated from operating' (Sorge *et al.* 1983, 150). This as it stands is not an adequate explanation unless the result of the differentiation is that less skilled labour can now be used for machine operating. Then the total process would become cheaper and one can see the advantages to management in fragmenting the task. It would presumably not be worth splitting the task into two for small batches because the communication costs between programmer and operator, with the frequent interaction needed given the small batch size, would outweigh the savings made in wage costs from hiring someone semi-skilled to do the machining.

The relations described above between batch size, company size, and work organisation were, however, modified by the type of cutting process and machinery used in fairly obvious ways. Thus the more time it took to program a job, the more likely that programming and operating were differentiated. This is because, at the time this study was done, a machine could not be programmed for the next job while it was engaged in cutting. Thus to combine programming and operating would, it was believed, reduce machine utilisation. It did not, however, lead to a polarisation of skills in every case. With complex machinery like machining centres, where something like sixty tools may be used to machine a workpiece, and where the piece and the machining head can be rotated in a number of ways, separate programming was done—often by ex-machinists—and the machine would be operated by a skilled operator. Polarisation tended to occur more on simpler machines, which could be programmed off the shop floor and operated by those with less skill than would have been required for conventional machines doing work of that complexity. With advances in electronic control and programming, and also possibly in operator programming skill, it is likely that in future it will be feasible for many operators to program the next job on the machine while it is still cutting the present piece. Experimental machines with this capability, and which are deliberately designed for operator programming, are already being developed by a group under the direction of Professor Howard Rosenbrock at the University of Manchester Institute of Science and Technology.

The differences between countries, already described in the earlier Anglo-French-German study, were left intact by the introduction of new technology. The German companies distinguished less than the British ones between careers and departments for production management and production engineering, work planning and work execution functions. This meant that in the German companies programming was becoming a nucleus of integration, bringing together operators, planners, production engineers and managers, charge-hands, and foremen, while in Britain that work was more concentrated on planners and production engineers. Moreover, in Britain such work was distinctive of white collar status, programming being the domain of planners and production engineers, while in Germany blue collar craft workers were used and there were instances of rotation between programming in the office and working on the shop floor.

Finally, Sorge and his colleagues observed that the relationship between this new technology and work organisation was influenced by the prevailing socio-economic conditions. In particular, they felt that enterprises in both Britain and Germany were having to cater for increasingly small market niches rather than for homogeneous mass markets. This was coupled with a move towards the manufacture of increasingly complex parts, requiring geometrically more demanding cuts and an increase in the number of cuts of different types to be performed on any one piece. In addition, there are pressures on companies to keep a check on the growth of fixed working capital in stocks and intermediate stores (work-in-progress).

All these consideration imply smaller batch sizes and more frequent conversions from a small batch of one complex product to a small batch of another. This increased variability is not one which can be handled bureaucratically, claim Sorge et al., and they have the weight of organisation theory on their side: it implies flexibility, autonomy, and high skill at the operator level. Therefore, as they observed, companies were increasingly emphasising the conservation or increase of shop-floor skills, and recognising the merits of an approach which relied strongly on craft-worker skills. This, they noted, was particularly so in the German companies which would presumably have a comparative advantage over British firms in this respect anyway.

The Scandinavian experiments

Perhaps the most well-known experiment in work organisation is the Volvo car assembly plant at Kalmar in Sweden where the company has made some attempt to break away from the traditional mass-production assembly line. Much has been written about this plant, but many misconceptions remain, partly because a great deal of what has been written is itself misleading. One of the major misconceptions concerns the scale of the enterprise. Contrary to one popular view, it is not the case that at the Kalmar plant each group of workers builds up a complete car from a box of component parts. In fact the Kalmar plant deals with only one stage of the production process, that of final assembly of the various components into the painted body. Even this part of the production process is itself broken down into many different stages at Kalmar. Moreover, Kalmar contains but a small part of the total final assembly facilities within the Volvo enterprise. This is not to deny that Volvo has made a significant innovation in work organisation and production technology, but it is one evolutionary step and not a revolution.

Kalmar is a smallish (50,000 population) town in a largely rural area of Sweden on the east coast of the country about 300 km. from the main Volvo production facilities in the large industrial town of Gothenburg on the west coast. The decision to locate an assembly plant at Kalmar was made in 1971 and seems to have been based on the need to expand the production facilities. It is, however, also the case that at that time the company was facing very substantial difficulties in recruiting labour. The Swedish economy was booming, there was very little unemployment, and routine assembly work was proving very unattractive to indigenous Swedish workers. At that time about 40 per cent of production workers in assembly were non-Swedish, a situation the company found unattractive because of the costs of recruitment, training, and supervision. To move to a rural area and to move away in some measure from traditional routine assembly techniques was seen as a partial answer to these problems.

The plant at Kalmar began production in 1974 with a planned capacity of 60,000 cars per year with two-shift working. In fact there has never been two-shift working at Kalmar and, because of the downturn in the car market after the 1973 oil price rise, production has rarely reached 30,000 per year at that plant. In 1980, for example, Kalmar assembled 22,800 cars. This is only 8 per cent of total Volvo car production worldwide, and five times more cars are still assembled on traditional lines at Gothenburg. Nevertheless, Kalmar is important as an example of what can be done by changing the technology of traditional assembly work.

Final assembly is the stage in the car production process when the welded steel car body, having already been painted, is brought together with the engine and transmission, and all the components and trim are attached to it to make up the completed car. In a traditional plant the engine, transmission, and axles are placed onto 'the track'—a chain-driven series of platforms which move continuously at a fixed pace while the operators work on the cars—and the painted body is lowered onto them. As the car then goes down the line, past various work stations, all the components are fitted to the car in turn. It will typically take several hours for the car to go down the line, hundreds of workers will be directly involved in fitting the various bits and pieces, and each will have a cycle time of perhaps 60 seconds to do their allotted task.

For the Kalmar project the company set itself the objective:

to build a plant which, without reducing efficiency, enables the employees to:
work in groups
exchange jobs through rotation
vary their working rate during the day
feel identification with the product
feel responsibility for quality
influence their working environment
communicate freely. (Company Factsheet 1980.)

Two of the major innovations at Kalmar were the architecture of the plant and the production technology. The plant comprises a series of four interlocking six-sided buildings designed to provide each work group with its own entrance, changing room, coffee room, and assembly area in an attempt to create a series of small workshops for each of the work groups and 'to promote team spirit' (Company brochure).

The major production innovation at Kalmar has been to do away with the track. Instead, the bodies are loaded onto individual carriers, known as automated guided vehicles (AGVs), which travel around the assembly plant under their own power and are individually guided by signals sent out through wires in the floor from a bank of central computers. The carriers can also be controlled manually. Carriers of this type are used in the building of the FIAT Strada/Ritmo cars and will be familiar to those who watch British television commercials. These are derivatives from the AGVs developed for the Kalmar plant.

The assembly task is broken up into segments, each being performed by a group of operatives in one bay of the plant, and taking about 20 minutes. Within each segment it was originally planned that work could be organised in either of two ways. It was always intended that the more usual of the two would be what Volvo term 'straight-line assembly', where work in one team area is divided among four or five stations, placed one after the other in the direction of production flow. The workers operate in pairs, following a car from station to station and carrying out the entire work assignment belonging to their team. When the pair finish one car they go back to the beginning of their segment and start on another. The other assembly approach which was originally planned for but has since been abandoned (*Volvo 10 Years On*) was 'dock assembly', where the carrier would move sideways into a dock in an assembly area. There, the team's entire assembly assignment would be carried out on the stationary vehicle. Each dock was manned by two or three people, who could swap assignments between themselves. The job cycle is thus about 20 minutes for any group of operators in either straight-line or dock assembly.

The flexibility introduced into the system by the carriers means that each group is able to operate with a degree of autonomy. Although the carriers are normally fully under the control of the computer, there is the facility for the work group to give commands to the computer from one of 40 VDUs scattered throughout the plant. They thus have some control over the speed with which the AGV moves through their section, and for those sections where the carrier does not move for the full 20 minutes there is the opportunity to pace their work over that span. Each group is responsible for its own quality control, is treated by management as a team, and there are opportunities for job rotation throughout the team, though the rotated jobs exclude supervision and materials handling. However, trade unionists and some managers indicate that the system is limited in its flexibility. Partly this is because management has placed limitations on who can operate the

VDUs, and partly the layout of the plant is such that there is only space for four cars between each work station. This does not provide a big enough buffer stock to allow each group to vary its speed of working through the day to a significant enough degree.

It is difficult to evaluate these innovations. To the casual observer the layout of the plant and the flexibility introduced by the carriers are striking improvements over the traditional assembly plant with the track making its inexorable progress past every fixed work station, and Volvo claim that the increased cost of building the plant in this way is covered by improvements in quality and productivity. However, although absenteeism and turnover are lower, claim the company, than the average for Swedish industry as a whole, it must be borne in mind that the plant is in an area with a low level of industrialisation. There has, moreover, been a continuously rising trend of absenteeism since the plant opened. Also, no hard data appear to be available giving a detailed breakdown of costs and savings associated with this new form of technology and work organisation, and it appears that, although there is spare capacity at Kalmar in the form of a second, as yet unworked, shift, when the new 760 model car was introduced in 1982 a new assembly line using conventional technology was set up for it at Gothenburg. A possible conclusion to be drawn from this is that management can provide alternative more attractive forms of work organisation but these may be at higher cost. Management may be prepared to pay this higher cost when labour is scarce and has to be attracted from alternative employment, or when a union is making strong demands for a higher quality of working life.

A second, less publicised but in many ways more interesting and far-reaching, experiment in new forms of work organisation is that which has been conducted in the other Swedish car company, SAAB, at Trollhatten. The location of the innovation there is not in final assembly but in the body shop where the various panels of pressed steel are welded together to form the body. SAAB were faced with the same difficulties as Volvo in recruiting staff in the 1970s, and they too have moved to a system of group working with job rotation in many sections of the plant. They have also developed the use of 'self inspection' —

each group responsible for inspecting its own work—and giving responsibility to work groups for minor routine maintenance such as changing tips of welding guns and making minor adjustments. There has also been attention paid to the design of the automated welding equipment in order to remove particularly poor job design features, but we discuss these later when we compare SAAB with BL.

However, their most striking innovation is in the way work is now organised in the last stages of the production process in the body shop. The final stage before the body leaves the body shop for the paint shop is normally for the completed welded body to pass down a line of welders and grinders. At various stations a number of final spot and seam welds are made, and then power grinders are used to smooth down exposed welds and any high spots. At SAAB this stage is removed from the line in what is termed a 'line-out' system. Bodies come into the section on a conveyor and are directed automatically into a buffer store. From there, a body will be directed into a dock where a pair of workers perform all the functions previously described. This includes considerable inspection work, firstly to establish what grinding needs to be done, and then to ensure that the body is fit, after they have completed their welding and grinding, to go on into the paint shop. When they have finished their work on one body, they activate the conveyor which carries off the completed body and pulls in another from the buffer store. The work is therefore not paced by the track, and the job cycle time is extended from seconds to about 20 minutes.

The performance of this form of work organisation is interesting and impressive. The company had expected it to show a quantifiable benefit by reducing absenteeism and labour turnover, and by easing the production control difficulties caused by the inflexibilities of the line system. One of these is that of balancing the line—having just the right number of operatives for the various work stations on the line. Another is that of maintaining production stability, because, for example, a hold-up at one work station may stop the whole line. SAAB had estimated the costs and benefits of the line-out system and had calculated that the system would pay for itself within four years, with more than half of the savings coming from the increased production flexibility of the system and another 25 per cent coming from reduced labour turnover and absenteeism. In the event, absenteeism in the body plant actually rose by over 20 per cent over the seven-year period spanning the installation of the line-out system, though this was a smaller rise than in the press and the paint plants adjacent to the body shop. Turnover fell by over 60 per cent in the body plant in the same period, though it should be noted that unemployment was rising in Sweden and therefore turnover rates in SAAB as a whole were falling, though less fast than in the body shop. The major saving from the line-out system came from the enhanced production stability brought about by the release from the inflexibility of line-working the system brought. Overall the system appeared to be even more cost-effective than the estimates, with a pay-back period of 2.6 years. Though one always has to treat figures such as these with caution, it does seem that the line-out project had considerable success both in terms of a reduction in operator dissatisfaction, reflected in lower turnover and a less rapid increase in absenteeism than other parts of the plant, and in terms of very substantial savings for management because of the increased flexibility of production it gave them.

Body assembly of the BL Metro at Longbridge

Our final case study is one conducted by the author and colleagues at the BL plant in Longbridge where the company had built an entire new plant containing much microelectronics technology to build the body of the new Metro small car. Many of the details of this study are reported in Willman and Winch, with Francis and Snell (1984), and we describe in some detail the way in which the company introduced this technology in Chapter 9. Of relevance to our discussion here are two choices facing the company about new forms of work organisation associated with the new technology. The first was an attempt to break down demarcation between skilled trades doing maintenance work—the moving towards so-called 'two-trades maintenance'. The second was to do with the design of the machinery to increase job cycle times and to reduce the extent to which work would be machine paced.

Within BL, as with other UK car companies, there are a large number of different trade groups specialising in particular aspects of machinery maintenance and repair. These include millwrights, pipefitters, sheet metal workers, electricians (only some of whom have electronics expertise), and machine-tool fitters. Each group in the past has operated with generally accepted lines of demarcation so that fault-finding and repair of a particular machine may involve several trade groups and take some considerable time while the fault is correctly identified and the appropriate craftsworker brought to the machine. With the introduction of more complex automated machinery, the difficulties posed by such a high level of demarcation became worse, and management attempted to negotiate an arrangement whereby maintenance crafts would be grouped into two broad areas—mechanical and electrical—with the understanding that demarcation within each group would be minimised. They were not wholly successful in this, and the compromise reached was termed 'two-trades response'. This meant that all electrical faults would be attended to by the electricians, but that mechanical faults would be dealt with by the mechanical trade nominated to that machine. By prior agreement some machines would be deemed to be the responsibility of millwrights and the others reserved for machine-tool fitters. The pipefitters' main responsibilities were for aspects of the plant's infrastructure rather than for particular machines.

In other countries, notably West Germany whose history of having industry unions such as the Metalworkers' Union has meant less demarcation between crafts, a series of possibilities about the organisation of maintenance have been discussed. To what extent might companies buy in maintenance from the machinery suppliers; to what extent might workers become skilled at the repair and maintenance of particular machines rather than skilled at a particular aspect of a machine's functioning, rather like the situation already obtaining with the repair and maintenance of household equipment such as washing machines, or office equipment such as photocopiers; and to what extent might companies take over the training of skilled workers, giving them courses specifically suited to the requirements of their own production systems and certificating them within house? These have been possible questions to ask for some time. Their pertinence is much increased by the rapidly increasing capitalisation of the work-place both in the factory and the office, and by the greatly increased complexity of the machinery. If BL, with its strong union presence, is prepared to negotiate changes in the work organisation of maintenance when microelectronics-controlled machinery is first installed, and management makes some headway in breaking down the traditional craft structure, then presumably substantial changes are on their way in other companies.

The second point of interest here about the BL case is the possibilities opened up with the new technology for improved forms of work organisation. Two aspects bear examination, and both can be related to the SAAB case outlined earlier. As it happens, the automated machinery used to build the Metro body is very similar to that used at SAAB and comes from the same manufacturer. BL, at the time the Metro facilities were being planned, was operating an extensive system of participation with joint management–union committees at several levels in the company and related to particular plants and projects. One of these committees was specifically concerned with the introduction of the Metro, and members of that committee, management and unionists, visited a number of car plants belonging to other countries, including SAAB at Trollhatten. Not only had SAAB innovated with their use of the line-out system, they had also, as previously mentioned, used group working to operate the welding machinery, and they had also taken care, in the light of union concern about the matter, to minimise the extent of machine-paced short job cycle time working. One technological fix for this problem is to fit carousels onto the welding machines. Automated welding machines need to be fed with the various pressed sheet steel panels and sub-assemblies which are then welded together under automatic control. For a typical rate of working of these machines, panels need to be fed approximately every 50 seconds. If they were fed in manually, this would imply exactly the kind of task objected to by the unions—namely, a task paced by the automatic welder and with a cycle time of less than a minute. The solution adopted at SAAB was to fit carousels to each machine. These carousels hold about twenty to thirty panels and feed the machine automatically. The manual job thus reduces to filling the carousel with panels or sub-assemblies every 20 minutes or so, and is of course only loosely machine paced. There is an advantage to management too, in that it reduces the dependence of the machine on consistent human performance. However, the carousels are a substantial capital investment.

Although BL did attempt to operate a modified and sim-plified system of group working at the Metro plant, they did not, despite the visit to Trollhatten, either use the line-out system or make extensive use of carousels. Moreover, there is no evidence of any discussion between management and unions about the possibility of doing so. Given SAAB's very positive experience with line-out, it is surprising that BL did not con-sider emulating it in that it appeared to offer benefits both to management and workers. BL have used carousels, but only to a limited degree and, it would appear, only where they were particularly vital to insulate the biggest machines from human variability.

Conclusions

These cases demonstrate that the simple de-skilling thesis, re-jected at a theoretical level in Chapter 4, is also unsupported by the empirical evidence, and for the same reasons. In some instances, such as in the robot-welding case, jobs are de-liberately de-skilled as part of a management strategy to tighten control and reduce costs. In others, as in the Scandinavian case and the optical factory described by Wilkinson (1983), management have deliberately reduced the tightness of their control and attempted to enhance skill levels. In some instances, management have done this because they have believed that de-skilling is a counter-productive strategy in that it lowers morale, commitment, motivation, and optimum use of human resources. In others, they have done it because they have felt they have had to: perhaps they would have failed to attract and keep labour if the jobs had been too fragmented. And certain features of the job design in the Swedish car factories were incorporated explicitly because of trade union action, either at local level or, more generally, because legislation had been enacted to require management to provide jobs of a certain quality.[1]

We now, in the next two chapters, try to assess the range of opportunities in particular technological contexts, and the costs and benefits to the various parties of particular strategies.

---

[1] The Co-determination Law of 1977. This is described in more detail in Chapter 8.

6

Organisation, Control, and Technology: A New Conceptual Approach

Introduction

There are, as we have seen in the previous chapter, a range of different organisational possibilities with new technology. Up to now we have considered the effects of different organisational arrangements on the quality of individual jobs. We have also examined the argument that technical and organisational changes are influenced by a class battle between capital and labour. However, the very existence of large-scale enterprises owes much to the fact that most modern goods and services cannot be provided by individuals working alone. They are the product of co-operative activities. These co-operative activities have to be co-ordinated, and a major function for any enterprise's organisation structure is to enable that co-ordination to happen. The form of organisation depends, to some extent, on the work the enterprise does and, hence, the kind of co-ordination required. As technology changes, so the co-ordination requirements change, and this is likely to lead to changes in the way work is organised. There are, then, a number of influences on work organisation. There is the pressure to create jobs which people find satisfactory, the pressure from management to increase the exploitation of labour, and the need to co-ordinate all the various jobs contained within modern large-scale enterprises. Technological change is likely to affect the influence which each of these pressures may have on the way work is organised. These next two chapters set out an approach to analysing the possible effects new technology may have on this complex process. The usefulness of the proposed approach is not that it will provide design solutions (because the choice of solution is a political process and will depend on various people's preferences and power bases), but that it may aid discussion among managers, trade unionists, and academics about possibilities for new forms of work organisation in the face of changes in process or product technology.

Technology, organisation, and control

In the pre-industrialisation era many goods and services were provided by single individuals. The village blacksmith, the carpenter, and the shoemaker worked on their own account, with the product of their particular craft or trade being sold direct to the final consumer. As goods and services have become more complex, this mode of operation is found much less frequently. No one individual could design and manufacture an entire automobile, for example. Cars contain many different products produced by entirely separate industries—a variety of parts, components, and sub-assemblies made from cast and pressed metals, plastics, and man-made fabrics. Each model will have been designed by various groups of people, and assembled by others out of components manufactured by yet others. The picture is similar for most modern consumer and capital products. Many services, and the fast food market with its strongly vertically integrated product line is an appropriate example, are provided via a similarly elaborate division of labour. Part of the reason for this elaborate division of labour is, of course, the process discussed in some detail in Chapter 4 whereby management attempt to wrest some advantage over labour by setting up such an arrangement. Though debate continues about the relative importance of this factor in explaining large-scale organisation,[1] it is generally accepted that there are some technical reasons for an elaborated division of labour. Large-scale tasks, requiring the simultaneous effort of many workers; complex products or services requiring a range of skills; and goods or services required in a hurry are all more efficiently provided by numbers of people, with different skills, operating collectively. If the production of a good or service necessarily requires the work of several people then interesting puzzles arise as to how the work is to be divided between these people, and how their various efforts are to be co-ordinated. Our interest in this question arises from our belief that, without understanding the processes going on in industry that establish a particular division of labour and form of co-ordination between that divided labour, we will not understand the demands that come from industry for particular types of technological de-

[1] See, for example, Francis (1983).

velopment, nor will we understand the relationship between these new technologies and the ways in which forms of work organisation are changed as these new technologies are introduced.

Both conventional organisation theory and the radical critique have a great deal to say about various methods of co-ordinating and controlling the activities of those performing a collective task. In this chapter we suggest that there is in this literature a concern about four basic issues over which control has to be exercised. Over some of these issues, different parties in the organisation will have conflicting preferences and be pursuing different outcomes. We then go on to examine the various strategies for control which have been suggested in the literature, relating these to the four basic control issues. Next we enumerate the various types of organisation usually considered by organisation theorists, and show how each type of organisation deals with the four basic control issues; we then tackle the vexed problem of the extent to which various organisational forms enable capital to exploit labour by examining the employment relationship and seeing how this fits in with our three typologies, of control issues, control strategies, and organisational types. Finally, we relate all this conceptualisation to our central concern over the relationship between work, work organisation, and new technology.

The concept of control

It is clear that the term 'control' is used in the organisational literature in a number of different ways, sometimes precisely, but usually very loosely. Braverman, for example, uses the terms 'control' and 'management' indiscriminately. Each involves the capitalists 'impos[ing] their will upon their workers . . . it involves the control of refractory masses'(1974, 67). 'Like a rider who uses reins, bridle, spurs, carrot, whip and training from birth to impose his will, the capitalist strives, through management, to *control*. And control is indeed the central concept of all management systems . . . '(1974, 68). In this sense the concept includes the classic sociological conception of power, defined by Dahl (1957) in the terms 'A [in this case management] has power over B [labour] to the extent that he can get B to do something that B would not otherwise do'.

This use of the term 'control' can be contrasted to its conceptualisation in systems theory. It is in this latter sense that Woodward uses the term, as we have seen in Chapter 3. The notion here is of a control system comprising four elements— objective setting, planning, execution, and control, where the control function is the monitoring of output, comparing it with the objective desired and adjusting variables within the execution process if output is not that which is desired (Woodward 1970). Control in this sense is termed 'co-ordination' by Braverman, who acknowledges that 'functions of management were brought into being by the very practice of co-operative labour. Even an assemblage of independently practising artisans requires co-ordination . . . ' (1974, 59). Much of conventional organisation theory is concerned with the exercise of control in this sense.

This latter sense in which the term 'control' is used contains the notion that individuals performing tasks which contribute to some larger end-product or end-service will be prepared to adjust their behaviour in order to enhance their contribution to the overall task. For example, players in a football team are constantly shifting position in the light of what they perceive to be the best interests of the team performance. Such co-ordination of activity may take place through what Mintzberg (1979) terms 'mutual adjustment', whereby individuals change their behaviour on their own initiative in the light of their knowledge of what other team members are doing, or it may take place as a result of specific instructions from the team captain. Continuing with the football team example, in most instances the individual player will have no personal preference for one action rather than another—is indifferent, as the economists would say. In these circumstances the control problem for the captain is that of gathering together enough knowledge about what is going on in order to produce instructions to each individual about what needs to be done. There is no question of the captain exercising power, in the sense in which we have just defined that term. For the purposes of this chapter we will label control in this sense 'Control 1'. Some degree of Control 1 will be necessary, as Braverman has observed, even among co-operating workers who have freely engaged in a contract between themselves, and possibly with an entrepreneur, to deliver an output in return for some specified reward.

Much of conventional organisation theory assumes that enterprises are made up of individuals who have entered freely into a contract to give their services to the enterprise in return for a given wage or salary. They are assumed to be indifferent, within limits which may be specified, to the tasks they perform. In other words, they do not mind what they are told to do. This is the authority relationship as Simon (1957) defines it. Most organisation theory concerns itself with ascertaining the most efficient mechanisms for processing information which will lead to the most appropriate instructions being given to these employees. Those who have undertaken more sociologically oriented studies of enterprises have often made the opposite assumption, that workers are not indifferent to what they do and that they prefer doing some tasks rather than others. Because of this, they are likely to attempt to do things the way they themselves prefer, and are thus likely to be in a power struggle with management over certain issues. Within this tradition, the exercise of power has been analysed at a level no higher than that of the enterprise, however, and so the radical critique, focusing on the conflict between labour and capital as social classes introduces a new dimension to the analysis of power within organisations.

What is necessary for an assessment of the radical attack on organisation theory is a conceptual framework for the analysis of control and co-ordination that will enable us to distinguish those features of an organisation which are necessary for efficient co-ordination of those who are indifferent in their preferences from those features which are the result of one group, or class, exerting power over others. It is to this end that we are suggesting that there are four major control issues facing any form of organisation, and that the radical attack can best be understood and evaluated on the basis of how each of these four control issues is solved. The first control issue is the one described above, that of co-ordinating different elements of a complex task performed by various individuals within some form of a division of labour when the individuals are indifferent between the range of alternative actions they might be instructed to engage in. We have labelled this 'Control 1'. This is the only control issue that is entirely free of the exercise of power or influence.

The three other control issues are to do with the degree of effort exerted by those in the collectivity. 'Control 2', as we term it, is about coping with the free-rider problem. The essence of this problem is that in any collective action any one individual is likely to gain a personal advantage by exerting less effort than the others, as the effect of one person's lack of effort on the corporate outcome is spread across all in the collectivity. Those who are not members of a trade union are commonly cited as examples of free-riders. Unless performing the work for its own sake is highly valued by the individual, or the organisation is strongly committed to extreme libertarian values, some form of Control 2 is likely to be valued by *all* workers, assuming everyone believes the majority of their number will be tempted to free-ride, but the exercise of Control 2 may nevertheless involve the collectivity exercising power, in Dahl's sense, over the individual worker.

Our last two control types are each concerned with control over the effort–reward bargain. We have already noted that conventional organisation theory often makes the heroic assumption about the indifference of individuals between different courses of action. In many circumstances an individual, faced with the need to change their activity in order to co-ordinate that activity with others, will feel that there are costs and benefits associated with the change. Simplifying the situation somewhat and assuming that the costs and benefits can be summarised in terms of the new activity requiring more or less effort than the old, we can draw this problem into the more general case of these third and fourth control issues.

For the sake of argument, let us assume that a line plotting reward against effort can be drawn which represents at all points along its length a consensus over what is an equivalent reward for any given level of effort (Figure 6.1). Thus, for example, in Figure 6.1 if, at point A, $r_.$ was the reward given for effort $e_.$, then there would be general consensus that, at point B, $r_.$ was the commensurate reward for the amount of effort $e_.$. A fairly uncontentious use of this graph would be to represent the line P as a certain rate of pay per hour, assuming,

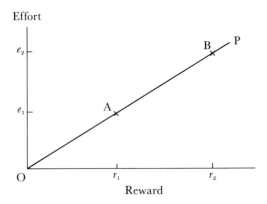

Fig. 6.1 The effort–reward curve

for the sake of this argument, that employer and workers all agree that it is equitable that, for example, those working ten hours per week should receive in total pay one-quarter the amount of those working forty hours. It is not, however, a necessary part of the argument that OP should be a straight line, nor that it should go through the origin.

We have already noted that individuals in altering what they are doing in order better to co-ordinate their work with others may perceive themselves to be altering their position on the effort–reward curve OP. The act of co-ordination may therefore involve reassessing rewards at the same time as issuing instructions to individuals to change their actions. However, the main reason for identifying the effort–reward curve as a major control issue is because it is conflict of effort and reward which the radicals have taken as their main point of attack on conventional views about work organisation. The Braverman thesis is concerned primarily with the proposition that capital uses particular forms of work organisation and technology to shift the slope of the effort–reward curve OP anti-clockwise, and the Marglin argument has as one of its major components the assertion that the rise of factory production, and the development of technology appropriate to that form of production, was due to capitalists' desire to force workers to operate at a point higher up the curve OP than they would prefer. The battle over the slope of the curve OP is our third control issue, 'Control 3', and over the position on the curve

OP 'Control 4'. We explore the implications of these two hypotheses later in the next section.

Preferences of capital and labour within each control type

There are, of course, many different interest groups associated with any organisation. Alan Fox (1971) uses the term 'stakeholders' to describe them. These stakeholders include the customers, owners of capital employed in the enterprise, suppliers, as well as managers and workers. Managers and workers as groups are also composed of sub-groups with different, and possibly competing, interests. Semi-skilled workers are likely often to have different interests from skilled workers, for example over demarcation issues. One department in the enterprise may have conflicting interests with another: for example, the sales/marketing function, with their likely concern for maximising sales, may well wish to pursue orders that lead to smaller batch production to specifications unique to individual customers, whereas the production department may well prefer the ease of producing goods with long production runs and to a standard specification. Such differences in interest are already fairly well discussed in the conventional organisation theory literature. The special contribution made by the radical perspective is to highlight the conflict of interest between labour and capital. For this reason, we now look in some detail at the supposed preferences of capital and labour over each control issue.

At first sight, our first two control issues—co-ordination (Control 1) and control over free-riders (Control 2)—may appear to give rise to relatively little conflict between capital and labour. However, we argue below that in all four types of control different preferences can be presumed.

Within Control 1, co-ordination of a complex division of labour, there is what Williamson (1975) terms 'atmosphere' to take into consideration. There may be conditions, for example when the task is rather complex and individuals' work is sequentially interdependent (Thompson 1967), for which the most efficient form of organisation might be mechanistic (to use Burns and Stalker's (1961) term), with much use of hierarchical management authority and little individual discretion. Where the 'atmosphere' of this type of organisation is disliked by any of the members, there is a potential trade-off between the benefits of high output and the costs of the 'atmosphere' created by the form of organisation yielding this output. There may be those who would be willing to forgo some of the benefits of high output for the preferred 'atmosphere' of a looser, more egalitarian form of organisation. Others, though, may view as a cost, not a benefit, the presence of organisational slack, putting a high value on efficient co-ordination and disliking the muddle and waste associated with slack. If workers are able to express their preferences freely, one would expect, *ceteris paribus*, to find a variety of organisational forms for any given technology, operating with varying degrees of efficiency, though with similar labour costs as workers trade productivity against atmosphere. Where, however, there is no direct connection between the overall output of the organisation and the rewards to individual workers, it does of course make no sense to postulate trade-offs over which individuals or groups of workers can exercise a degree of choice.

Control 2 displays a similar trade-off. When work is being done by a small group of peers, control over the free-rider can usually be exerted by informal social controls, though even in these circumstances not always very effectively—as members of university academic departments can sometimes testify! If an enterprise is large, however, some elements of formal organisation are even more likely to be necessary to meter individual effort and to sanction free-riding. There may thus be a trade-off within Control 2 between efficiency in preventing free-riding and 'atmospheric' conditions relating to individual freedom. As with the trade-off in Control 1, one would expect individuals to differ in their preferences with regard to the point of balance between freedom and efficiency, and one might expect to find a variety of organisational arrangements to match, if individuals had the freedom to choose organisational forms to meet their preferences. The radical attack does not concern itself with these issues.

One of the two major foci of the radical attack is conflict over the effort–reward bargain, the issue we have already labelled 'Control 3'. The simple case of this conflict is that which is assumed to operate within the employment relationship, where the employee exerts the effort and the employer provides the reward. The two parties in this conflict are usually regarded by the radicals as belonging to the two classes of labour and capital, though in the case of publicly owned institutions this may not necessarily be the case. Within this simple case, the control issue can be conceptualised from Figure 6.1 as conflict over the position of the line OP. Many years ago, Baldamus (1961) introduced the same model[2] at the end of his study of efficiency and effort. If the two parties to this conflict are characterised as capital and labour then, *ceteris paribus*, capital will prefer P, in Figure 6.1, to be as near to the vertical axis as possible, so that the effort–reward ratio is high, whereas labour will prefer P to be near the horizontal axis. Control 3 is the exercise of power that managers can exert over labour to push the line in the preferred direction, and it is by the use of Control 3 that capital may exploit labour in the classic Marxist sense of extracting surplus value.

The non-exploitative position of the line OP would, for the Marxists, be where the value of the labourer's work was appropriated in its entirety by the labourer. A shift in OP anticlockwise represents surplus value being extracted by the capitalist. A shift clockwise represents the capitalist subsidising the labourer. In terms of conventional economics, the equilibrium position of the line OP would represent the wage rate for a given level of effort that would be dictated by the market as employer and employee freely entered into a contract for a particular piece of work. It would result in the pay to the employee being equivalent to the marginal productivity of labour. Deviations from that line would come about owing to a degree of monopoly power residing in either party to the contract.

The moral aspects of Control 3 have been extensively discussed by labour economists (for a review see, for example, Willman 1982). However, the issue over the effort–reward bargain that is the subject of Control 3 goes rather wider than the arena of conflict discussed by the radicals. One possibility that should be pointed out at this point is that, although the radicals usually highlight the exploitation of workers by managers,

[2] I am indebted to Richard K. Brown for drawing this to my attention.

there may be occasions when, if workers and the employing organisation or both are in monopoly positions, employees in one organisation may be exploiting other workers by bidding up the wage rate in one company/industry, with the high wage being passed in high prices to other workers as consumers.

There may also be conflict, it should be noted, over what constitutes an equitable slope of OP between different occupational groups in any collectivity—even if that collectivity is a producer co-operative. Different groups may dispute appropriate methods of measuring effort, of assessing what rewards, both of a non-pecuniary and pecuniary kind, are gained, and what is an appropriate relationship between these for any one occupational group. Debates over what constitutes an equitable teaching load within university departments provide a classic example of this aspect of control.

The final, and perhaps most important, extension of the Control 3 issue is to include aspects of co-ordination control within this category. Under the label 'Control 1' we include co-ordination of individuals involved in collective activity in so far as those individuals were indifferent between different activities. In many cases, however, an individual will experience a particular basket of costs and benefits associated with any one task. Performing a different task will result in a different bundle of rewards. Therefore the act of co-ordination, where it involves individuals being asked to alter what they would have been doing, is likely also to involve the co-ordinator in assessing whether this change is also affecting that individual's position on the curve OP. Thus much of Control 1 is likely to involve Control 3 as well.

Let us now turn to what is at the heart of the radical attack, at least as far as its implications for work organisation and new technology are concerned.

Much, but by no means all, of the radical approach to organisation theory has been concerned with ways in which organisations are developed and used by employers and managers to swing the line OP anti-clockwise. Four main strategies are described in the literature.

One is that of breaking union solidarity and thus shifting the relationship between capital and labour from duopoly to monopoly. This may happen, as in the case of the iron and steel industry described by Stone (1973), without management changing the skill composition of the labour force, or may involve the substitution of skilled by less skilled workers. Without the backing of a solidaristic craft union, less skilled workers may be more easily sacked and the sense of insecurity, Stone argues, makes them submit the more easily to management demands for high levels of effort.

The second managerial strategy for shifting the line OP anticlockwise also involves de-skilling, but is based on a different logic. It is:

That the master manufacturer, by dividing the work to be executed into different processes, each requiring different degrees of skill or of force, can purchase exactly that precise quantity of both which is necessary for each process; whereas, if the whole work were executed by one workman, that person must possess sufficient skill to perform the most difficult, and sufficient strength to execute the most laborious, of the operations into which the art is divided. (Babbage 1963, 175–6; quoted by Braverman 1974, 79–80.)

This shifting of the line OP by labour cheapening through elaboration of the division of labour increases the efficiency of the operation only in the sense that labour *costs* are reduced. The labour effort remains the same.

Whether this strategy represents increased efficiency or the power of capital depends, therefore, on one's view of the operation of the labour market. Braverman claims that 'the capitalist mode of production creates a working population suitable to its needs' (1974, 82) and thus directly influences the shape of the labour market. Others would doubtless argue that the labour market is, in part at least, a reflection of variation in the abilities of sellers.

The third managerial strategy is the one to which Braverman devotes most attention—that by which management through the use of 'scientific management' obtains knowledge from workers about maximum possible levels of output and renegotiates, or forces, a much higher level of output for rather less of an increase in pay. Taylorism, of course, also enables management to pursue the de-skilling strategy discussed above.

The fourth strategy is one of those analysed by Marglin (1974), in his description of the shift from putting-out to factory work in the eighteenth century in Britain. One reason for the merchants' shifting production in this way was the tighter control they could exercise over labour in the factory by employing supervisors to 'drive' labour. Some of the extra effort for a given reward came, he claims, from the tighter discipline managers had over operators within the factory.

Much of the thrust of Marglin's argument is, however, directed at the question of Control 4, the issue of where *on* the line OP an individual worker will operate. A and B in Figure 6.1 may represent effort–reward ratios which are in some sense regarded by both employer and employee as equally fair, but B produces more output (for more reward) than A, and employers may still have a strong preference for workers to operate at B rather than A because, for example, there may be a set of relatively fixed costs per worker. Marglin's argument in relation to the setting-up of factories was that workers had such a strong preference for leisure that beyond a certain point on the curve OP they would refuse to work if given any choice. Labour economists today recognise that many workers exhibit a strong preference for leisure rather than extra income beyond a certain point, so that in terms of Figure 6.1 they may prefer to operate at a point nearer to A. Given the possibility that employers will prefer to operate nearer to B, then Control 4 is the power which employers/managers attempt to exert over workers to shift them *up* the line OP. Note that, although it may be against the preferences of some workers to be pushed up the line, it is not increasing the degree of exploitation in the Marxist sense of extracting more surplus value for a given amount of effort. Note also that on occasions it might suit management to attempt to shift operations back down the line OP—for example, in slowing down production, or even triggering industrial action, when stock levels of the finished product are running at higher levels than seem justified by sales forecasts. Control 4 may involve the exercise of power at two levels. The superficial, though real, level is that where power is being exercised by management to force workers to perform more strenuously, or for longer hours, though for more money, than those particular workers would prefer. In other words, the workers have a higher preference for leisure than management wishes them to exercise. Such a preference may be exercised through absenteeism or by low levels of productivity at work in the face of incentive payment systems, for example. Where management is exercising tight Control 4 one would expect to see punitive controls over absenteeism, and non-reward-linked methods of ensuring high labour productivity while workers were on the job.

The deeper, more profound level at which Control 4 may operate is in the extent to which worker preferences are shifted to the high reward/high output position by various mechanisms which owners of capital may have to hand—the power of advertising being the most obvious. There are those who would introduce Gramsci's notion of cultural hegemony at this point, too (see Clegg and Dunkerley 1980, 492–501, for example).

Three types of organisation and their control characteristics

Having established four primary issues over which control may need to be exercised over the pursuit of collective action and the interests of some of the more important groups involved, we can now move on to the main purpose of this chapter, which is to explore the implications of these control issues for the way work is organised. Not until we have done that are we in a position to establish any possible linkages between new technology and work organisation.

In order to pursue the question of what are the capacities of various forms of organisation to enable various groups of organisational participants to exercise control over the various issues, we set up a typology of organisational forms. In order to establish what Abrahamson (1982) calls the logic of the situation, we assume for this next section the absence of an employment relationship. We do this in order to build up our analysis one step at a time. If we begin the analysis by assuming a group of equals attempting to co-ordinate their activities by various means without any outside intervention, we can explore some of the basic organisational processes through which control relating to our four issues can be exercised. Obvious examples of such groups are producer co-operatives, voluntary associations of various kinds, and groups of working professionals such as lawyers or civil engineers who organise their practises by means of partnerships. Only after we have established how control may be exercised by means of various organisational forms where there is an absence of the employment relationship do we then discuss the more common, but more complex, situation where many of those whose activities are being co-ordinated are employees standing in an authority relationship to their employer. We will use the term 'entrepreneur' rather than 'capitalist' for the employer, both because this sounds more neutral and also, as we discuss later in this chapter, because the entrepreneur may be hiring capital as well as labour. We will reserve the term 'manager' for someone who is exercising co-ordination, whether or not those being co-ordinated by the manager are employees or fellow members of, for example, a producer co-operative.

If, then, we consider how work might be organised in order that there may be effective control over the issues we have delineated, what are the major forms of organisation structure we might envisage? We follow Williamson (1975) for this purpose, and use the categories of market, peer-group co-ordination, and hierarchy of various forms. That market relationships are a form of, rather than an alternative to, organisation is an idea which is foreign to most non-economists. Pugh *et al.* (1983), for example, in their summary of Williamson's contribution to organisation theory continually set up market and organisation as alternatives. Economists, such as Williamson, contrast market not with organisation but with hierarchy. Their argument is that the co-ordination of collective activity in the production of some good or service can be organised either by the various parties buying and selling the products of their sub-tasks along the production chain, or by their joining together and co-ordinating their activities either within a peer group or by setting up a managerial hierarchy of authority.

Let us go back to our example of car production to illustrate the point. A modern motor vehicle is made up of very many different components, and many components go through a number of stages in the production process. An inlet manifold (a component part of the engine) may well be made out of cast metal. To make the manifold, molten metal is poured into a die. The die itself is machined by skilled machine-tool operators, to specifications laid down on an engineering drawing drawn up by a detail draughtsman. That draughtsman will have drawn the manifold to a particular shape on the basis of design work done by professional engineers as a result of aerodynamic research carried out by yet other engineer. It is possible that this whole process of research, design, drawing, machining, and casting may take place within the one company, and relations between the engineers, draughtsmen, technicians, machinists, and foundry workers would in that case be co-ordinated by a managerial hierarchy. It is much more likely, however, that the basic research was done by a university professor (or the professor's research assistant), and the design and detail drawing done by the motor manufacturer, while the die may well have been machined by a small, independent, high-precision engineering company, and the casting done by yet another independent foundry company. If this were the case then much of the co-ordination of the various activities needed to produce the inlet manifolds would have taken place through the market place as the various companies entered into tenders and contracts with each other. In each case the various actions are organised. In one case, the organisation is by an administrative hierarchy. In the other, it is by the market.

A question to which researchers are now paying considerable attention is under what circumstances the one form of co-ordination is more favourable than the other: most of the rest of this chapter is devoted to this question. The reason for exploring this question here is twofold. Firstly the form of organisation which is chosen to co-ordinate a set of activities will have implications for the way in which new technology is developed and used—particularly if we are concerned with the use of information technology. Secondly, the developing technology may well shift the costs and benefits of using a particular form of co-ordination. We may thus see that, with the coming of new technology, firms will shift from using market to hierarchy or vice versa as their preferred form of co-ordination for particular transactions. We explore this particular point in Chapter 7.

Returning to our original argument, we are concerned at the moment to investigate the properties of market, peer group, and hierarchy as alternative forms of organisation for exercising control. Williamson, as an economist, naturally argues that market as a form of organisation is logically, and usually temporally, prior to peer group or hierarchy, but this is not a necessary part of his argument. Marglin is, of course, describing in his analysis of the development of the British textile industry just such a transition from market to hierarchy, though he does not accept Williamson's explanation (Francis 1983).

The peer group may adopt a variety of forms of organisation, but its defining characteristic is the absence of an overarching authority structure. There are, of course, a number of ways of co-ordinating work among equals, without setting up a managerial hierarchy, which go beyond simple face-to-face discussion; indeed, most forms of collective activity use them in addition to the use of a managerial hierarchy. Mintzberg (1979) gives a fairly exhaustive list of such co-ordination devices. There might, for example, be mutual agreement between members of the peer group, in advance of action, over rules and procedures to be followed. Most voluntary organisations, for example the local tennis club, will have these. There might, in addition, be agreement in advance over what should be the standardised outputs produced by each worker. Suburban baby-sitting circles, or rotas to run the scout jumble sale, provide homely examples of these. There could also, if tasks were rather complex, be a form of standardisation of skills through a training process. Professional practices, such as a solicitors' partnership, utilise this form of co-ordination. All these are additional to the particularly common form of co-ordination—through mutual adjustment. Any collectivity, whether a voluntary organisation, a professional partnership, or a producer co-operative, is likely to use some combination of these techniques. Nevertheless, in most collectivities, particularly those in which there are a diverse range of occupations engaged in a single task, some form of hierarchy is usually adopted. What follows is a discussion of the various ways in which markets, peer groups, and hierarchies cope with our four basic issues over which control is exercised.

Table 6.1 is an attempt to illustrate this discussion. It is in the form of a three-by-four matrix, the horizontal dimension being the three types of organisation, while the vertical dimension sets out the four control types. Each box in the matrix summarises how that form of organisation—market, peer

TABLE 6.1

The relationship between control issues and organisational type

Type of control	Type of Organisation		
	Market (co-ordination and control via price) *tenders/ contracts*	Peer group (co-ordination and control via rules, procedures, standardisation, mutual adjustment)	Hierarchy (co-ordination and control via hierarchy, planning, product organisation, matrix)
Control 1 (co-ordination)	Suffers from defects as identified by agency theory and TCA	Unable to cope with large-scale co-ordination	Particular type of arrangement as predicted by contingency theory
Control 2 (free-riding)	Problem when one party is in a monopoly position	Problem of metering individual effort	Allows metering
Control 3 (effort–reward bargain)	Suffers from defects as identified by agency theory and TCA, and when one party is in a monopoly position	Okay so long as group can operate social control	Allows an employment relationship
Control 4 (position on effort–reward curve)	Problem when one party is in a monopoly position	Problem of metering individual effort	Allows metering

Note: TCA=transaction-costs analysis.

group, or hierarchy—copes with the particular control issue. In the discussion that follows we deal with each of the three forms of co-ordination in turn.

In the market mode of organisation all four types of control are exercised by the price mechanism. In the pure case activities between separate business units are co-ordinated by means of price signals in the market-place. Control 2 is not an issue because individuals set a price for their own activity. The effort–reward bargain (the Control 3 issue) is struck under conditions of perfect competition in a free market, and each is free to choose what position they want to operate on the effort–reward curve (Control 4). However, Control types 2, 3, and 4 all suffer problems if one party is in a monopolistic position, or there are other reasons why competition is not perfect or markets free. These problems are familiar to all economists and handled by conventional economic theory. Moreover, recent economic developments in agency theory and transaction cost analysis (TCA) have identified a series of conditions which, if present, prevent the market mode of co-ordination exercising effective control over free-riding or the effort–reward bargain—Control 2 and 3.[3] It is also possible that the information impactedness and opportunism resulting from the circumstances causing the break-down of Control 2 and 3 may lead on to a breakdown in Control 1 because incorrect price signals will be transmitted. When producers join together and decide to co-ordinate activities through peer-group organisation, Control 1 may be exercised through the methods already mentioned—rules, procedures, standardisation of output or skill, and mutual adjustment. Control 2, in small peer groups, might be exercised through social controls of the kind familiar to social psychologists. Within peer groups, if they are producer co-operatives, the Control 3 problem is limited to the internal allocation of that income, gained by the group as a whole, obtained either from selling the good or service in the market place or from a budget allocation if the good or service is a non-traded commodity. The internal allocation has to take account of the problems already discussed in relation to Control 3 about appropriate measures of effort and reward and their ratio for different occupational groups; it also may have to address the Control 4 problem—namely, that some members of the peer group may wish to operate at different points on the curve OP (see Figure 6.1). Both these problems may therefore raise the issue of metering effort. Dislike of intensive metering, and the difficulties of performing metering in a peer group, coupled with poor opportunities for exercising social control among peers, may lead to peer-group organisation being less favoured. Though the form of co-ordination it offers for Control 1 may be effective, its ineffectiveness in handling

---

[3] For an exposition of agency theory see, for example, Alchian and Demsetz (1972), and of TCA see Williamson (1985).

Control types 2, 3, and 4 may lead to market or hierarchical organisation being adopted instead.

Hierarchical organisation has as its fundamental element the authority relationship. This, as Weber (1947) defmes it, is a relationship where one person, A, willingly allows another, B, to direct him or her to do something which he/she would not otherwise do. In the rational–legal case (the one conventionally deemed to exist in most 'work' enterprises) A considers B's authority over him/her to be legitimate either because A accepts the overall goals of the collectivity and B is deemed to be directing A in such a way as to maximise the achievement of these goals, or because A is indifferent to the overall goals but is prepared to accept B's authority as part of what he/she is paid to do.

It is usually argued by conventional organisation theorists (and particularly rigorously by both Mintzberg and Williamson) that co-ordination by hierarchy is more efficient than peer-group co-ordination for the co-ordination of complex tasks which involve an elaborate division of labour and high levels of uncertainty. They argue that in these circumstances rules, standardisation, and mutual adjustment through a fully coupled network are ineffective and inefficient.

Hierarchy may also offer an advantage over peer group for solving control problems of types 2 and 4 because it provides a mechanism for metering individual effort. If different members of the collective have different preferences for the position on the line OP, hierarchy may provide a method of giving requisite variety, as individuals may be able to chose their positions on the line and accept the metering of effort required to allow rewards to be distributed equitably.

It would be argued by many that even producer-owned co-operatives might therefore choose to use a degree of hierarchy in their organisations because of its efficiency advantages. Positions in the hierarchy could be filled on a rotating basis or by democratic election either for a limited period or for life. Nevertheless, some may prefer the 'atmosphere' of a non-hierarchy, and, though hierarchy may bring with it higher productivity, they may prefer to trade this off against the preferred atmosphere of the peer group.

A number of economists have pointed to the existence of

internal labour markets in organisations (see, for example, Do-eringer and Piore 1971, and Stone 1973), and the theory of such markets has been extended by Williamson (1975). The principal notion is that entry into an organisation is only at a low level, more senior positions being filled by internal promotion only. The significant feature of internal labour markets for a discussion of the operation of hierarchy is that tasks which could be done by one person may, where internal labour markets operate, be fragmented into a hierarchy of sub-tasks, the more simple elements being performed by those at lower levels in the organisation and the more demanding elements given to those who have worked their way up the job ladder in the organisation. The internal labour market, it is argued, is an efficient way of coping with 'circumstances where workers *acquire*, during the course of their [working life in that particular collectivity], significant job-specific skills and related task-specific knowledge' (Williamson 1975, 57). It copes with the problem posed by the temptation of workers with these idio-syncratic skills to act opportunistically (the Control 2 problem), and also allows those who wish to develop such skills some reward for so doing (the Control 4 problem). Whether internal labour markets come about by some deliberate manipulation of job content, or result from some given characteristic of the technology, it is conceivable that even a producer co-operative may use some form of internal labour market to solve Control 2 and Control 4 problems.

We have not, so far, discussed the use of hierarchy in connection with the Control 3 issue—the classic exploitation sense of control. This is because we have not yet discussed the employment relationship specifically. Without an employment relationship, and an entrepreneur, the Control 3 problem only arises in the limited sense already discussed. It is the presence of an entrepreneur that creates the condition for the exercise of Control 3, because the person performing a particular sub-task within a collectivity no longer shares the reward to the collectivity from performing the collection of tasks which create that collectivity's output good or service. That reward goes to the entrepreneur, who passes on a proportion of it to the individual worker who stands in an employment relationship to the entrepreneur.

Interestingly, Williamson (1975) moves on from a discussion of peer group and simple hierarchy, where principles applicable to all collectivities are enunciated, to a discussion of internal labour markets and the employment relationship, without discussing the implications of the presence of an entrepreneur. 'Capitalist' organisation is assumed, with an entrepreneur either owning or hiring capital and also hiring labour to produce some good or service which the entrepreneur sells. The possibility that this arrangement could be stood on its head, and labour could hire capital and entrepreneurial skill, is not discussed by Williamson, although some of the industrial relations implications of this alternative arrangement are explored by Willman (1983).

The argument so far is that there are a variety of control problems which need to be solved by some mix of organisational forms, whatever the ownership status of the enterprise. However, introducing the entrepreneur brings into the picture a further set of control issues which are likely to have significant organisational implications.

The employment relationship and the role of the entrepreneur

The discussion so far has been concerned with various forms of organisation without any consideration of the existence of an employment relationship. We have been concerned to show that types of organisation differ in their capacity to allow particular types of control to be exercised. However, most forms of work organisation, in the West at least, incorporate an explicit employment relationship, whereby labourers are hired to perform a more-or-less specified set of tasks in return for a more-or-less specified set of rewards. Often the employment relationship is expressed in some form of contract. For the purposes of the present discussion, the person hiring the labour will be called the entrepreneur. One question is why the usual practice is for entrepreneurs to hire labour, rather than vice versa, and another is that of the relationship between the entrepreneur and capital. Does it make a difference to how control is exercised in an enterprise if the labourers hire the entrepreneur, rather than the entrepreneur hire labour, and does it make a difference if the entrepreneur hires capital rather than the entrepreneur being a capitalist, or an agent of capital? To pursue these questions we first suggest six possible functions performed by the entrepreneur in organising labour, and then pursue the implications of these functions for the debate about organisation and control.

1. *The entrepreneur as cognoscente.* Marglin (1984) suggests that the entrepreneur in the late eighteenth century in the textile industry brought to the industry new skills in organising production within a managerial hierarchy, and that the skill and effort involved in producing this new form of organisation produced extra output. Marglin argues that these skills and knowledge of organisation and management are technically public goods but that the *cognoscenti* were able to exercise some monopoly control over them and thus able to charge a rent for this knowledge and skill. McGuinness (1983) also suggests this interpretation of the entrepreneurial role, but accepts that the entrepreneur's skills may be private goods and thus legitimate to rent out. He also suggests that in some cases workers may gain more than they may lose in moving into an employment relationship with an entrepreneur. If this is the typical contribution made by an entrepreneur, one would expect to find this role being performed in a number of enterprises, including those such as voluntary or professional organisations where the workers hire the entrepreneur (and call him/her an administrator) rather than vice versa.

2. *The entrepreneur as hero of the unskilled.* There are, however, a clutch of functions which the organisational entrepreneur may perform which are more ambiguous in the desirability of their effects. These functions correspond to the various labour-controlling managerial strategies already discussed. Each involves the entrepreneur in adding managerial work to, or subtracting mental work from, the manual work of the workers; however, the intentions of the entrepreneur, the process of adding the mental work, the evaluation of the outcome by skilled and unskilled workers, and whether labourers would ever conceive it to be in their interests to hire an entrepreneur to carry out this work for them are different in each of the three cases now to be discussed.

The first of these three is the entrepreneur as the visible hand implementing the Babbage strategy (1963, described above) of dividing the work to be executed into different processes, each requiring different degrees of skill or of force. To the extent that this strategy reduces labour costs, there is a rent to be earned by an entrepreneur here. This is *not* always a strategy which operators will resist. If there is ample work for the highly skilled to do, even after some of the less skilled aspects of their tasks are taken from them, and there are less skilled people seeking work who are given new work opportunities by the implementation of this strategy, then reorganising work along these Babbage lines might under these circumstances be generally welcomed. Obvious examples are those of professional workers such as doctors, lawyers, and engineers who might choose to employ paramedicals, clerks, technicians, and even machines to do certain relatively routine aspects of their jobs. It is possible that in a significant number of instances those who are given these fragmented tasks falling from the professional table are rather grateful for them, and so the new arrangement may be in everyone's best interests. The radical critique would, of course, be that the subordinate semi-professions were suffering from false consciousness and also from a false view of their own capabilities. The preferable alternative, from the radical perspective, would, presumably, be one where all those involved in performing these collections of activities would be trained to the same level, and the esoteric and routine elements in the jobs be shared equally.

3. *The entrepreneur as trust-buster.* Another version of this entrepreneurial act would be that of carrying out the strategy, described by Stone (1973), of fragmenting jobs simply to break up the existing monopoly of a traditional craft organisation over a particular occupation. It is difficult to imagine labour employing someone with entrepreneurial skills to carry out this task for them, though it is possible to imagine a situation where other occupational groups in an enterprise might welcome an entrepreneur breaking the monopoly of one occupational group over a set of tasks.

4. *The entrepreneur as monopoliser of mental tasks.* It is unlikely that labour would be willingly involved in this fourth category of

entrepreneurial action. This is the Marglin (1984) interpretation of the entrepreneurial role. Not only does the entrepreneur make a contribution in installing a new form of organisation—one that splits manual from mental elements of the tasks—but he then goes on to appropriate the public good of this managerial knowledge and gain a rent from his monopoly hold over it. The entrepreneur's profits thus accrue at least as much from the power obtained from his monopoly as from the reduced costs of labour from the Babbage-type effects of fragmenting the task.

5. *The entrepreneur as broker between labour and capital.* Fifthly, the presence of an entrepreneur in the labour process may be explained in some measure and in some instances by that entrepreneur's access to capital and/or product markets. If such access results simply from entrepreneurial initiative or reasonably freely available management training and the development of a specialised competence about the technicalities of how capital and product markets work, then it might be regarded by workers as reasonable that someone should earn a rent from this entrepreneurial activity.

6. *The entrepreneur as capitalist.* The radical argument, however, would be that one's class position in a society is extremely influential in giving access to capital—and control over the means of production through control over the means of marketing the product. If this view is taken then the entrepreneur's access to capital, and his control over product markets, should more correctly be seen as resulting from the exercise of power and privilege rather than of skill.

The purpose of the foregoing analysis is similar to that of the analysis of the relative importance of power and efficiency in explaining forms of organisations. Our conclusion is that there are some functions performed by the entrepreneur which appear to be justified on efficiency grounds and which would be welcomed under certain conditions even by members of a producer co-operative. Other functions seem to involve the exercise of power and are likely to result from capital imposing its will on labour.

Our argument, therefore, is that even in the absence of power relations between capital and labour there are a variety of forms of organisation which a collectivity of workers may choose from. The principal types of organisation are market, peer group and hierarchy, with or without the presence of an entrepreneur and employment relationship. These will have differing capacities for co-ordinating tasks, for controlling free-riding, for determining the relationship between reward and effort, and for setting objectives about effort levels. Conventional organisation theory has in the past concentrated too much on just the first of these issues, leaving other sub-disciplines, particularly industrial relations, to address the latter three questions. We argue here that these too are organisational issues. We also argue that there may be a trade-off between efficiency and atmosphere, and that organisational forms will differ in the trade-off they provide. Individuals and groups may have different preferences for the trade-off they may wish to make and thus for any given level of task complexity (and indeed for any given set of conditions of size, technology, and environment usually considered by contingency theory). One ought to find a range of organisational forms reflecting these different preferences. If no such range exists, this may be because preferences are being overridden by some set of powerful interests.

In some circumstances the search for efficiency, or possibly better atmosphere, may lead a collectivity to hire an entrepreneur to manage the organisation. Such a choice is not unknown in voluntary organisations or among professional groups and does not necessarily lead to the members of that collectivity moving from the position of independent contractors to that of working within an employment relationship. Nevertheless, even within a group of self-employed professionals, or an independent producer co-operative, it is conceivable that the adopted form of organisation will have a degree of complexity, of managerial hierarchy, and of fragmented tasks that owes something to the management control considerations posed by the labour process school. In the terms of the concepts used here, the control problems 2, 3, and 4 (control over free-riding, over the effort–reward bargain, and over effort intensity) may well exist even in democratic work institutions and thus lead to organisational forms which the radical critique has identified as being explicitly 'capitalist'.

Nevertheless, we would go on to argue that, within capitalism the functions of the entrepreneur are extended to force an employment relationship onto workers so that capital may exercise power over labour particularly in order to swing the effort–reward bargain more in capital's favour, and also,but perhaps to a lesser extent, to push labour up the effort–reward curve in the direction of more intensive effort than labour would prefer to exert. To do this, labour is fragmented further; more managerial hierarchy is instituted in order to co-ordinate the more elaborate division of labour; informal social controls over free-riding break down and the hierarchy has to be elaborated still further to exercise control over this; and further organisational arrangements have to be instituted to ensure higher levels of effort.

Analysis of organisation therefore requires attention, not just to the task arrangements and division of labour at any given moment, but to the preferences of organisational members, the distribution of power, and the way in which the four control problems are being solved. However, as we suggested in the Introduction to this chapter, the usefulness of this conceptual framework lies less in its capacity to aid analysis of extant organisational forms than in its potential for helping discussion among academics, trade unionists, and managers about possibilities for new forms of work organisation in the face of changes in, for example, technology or product lines, or when issues of participation and industrial democracy are on the agenda. In the next chapter we use this conceptual framework to develop a discussion about some of the possibilities created by the development of new technology for these wider organisational issues.

7
Organisational Design and New Technology

Introduction

In the preceding chapter we set out a conceptual framework intended to aid our understanding of possible ways of organising work under various circumstances. The question now is that of the implications of new technology for the organisation of enterprises and their management.

A major conclusion from the analysis in the last chapter is that there are a large number of possible forms of organisation, with differing capacities for the exercise of control and with characteristics that are likely to be valued differently by competing groups associated with the enterprise. Thus the form of organisation which is adopted by an enterprise will, other things being equal, be the one which has the characteristics favoured by the most powerful group. Any prediction about the relationship between new technology and organisational form would therefore require foreknowledge of who were the various interested parties in the enterprise and how much power each had. Such knowledge will usually be specific to the enterprise. Moreover, the new technology is only now beginning to be in widespread use and it is not yet possible to discern any firm trends in patterns of usage. Thus the purpose of this chapter is not so much to make forecasts about the impact of new technology on organisational design, though we do attempt to sketch out what seem to be likely trends, but more to continue to enumerate the factors which are likely to influence those influential in the organisation design process. The specific features of management and organisation examined in this chapter are: firstly, the likely impact of new technology on managerial and technical functions within the enterprise; secondly, possible effects on the overall size of companies and work-places, including the question of telecommuting; and, thirdly, changes in the shape and complexity of organisation structures.

Changes in the nature of managerial and technical work

Changes in the managerial task

In common parlance there is much misuse of the term 'manager' because of the tendency in many enterprises to use the word to describe someone's status rather than their job. People are promoted into a 'management' grade though the job they do may be that of providing technical expertise. In our discussion here we reserve the term 'manager' for someone who has responsibility for the management of a function or group of people—that is, someone who sets objectives, organises the attainment of those objectives, and monitors, and is responsible for, the performance of that section of the enterprise. Such a person may also have professional or technical responsibilities of their own, but there are likely to be a number of people in any enterprise who have considerable technical expertise and responsibilities and who are of at least equivalent status to managers, but who have no managerial responsibility. What we want to discuss is the impact of new technology first on managerial work as we have just defined it, and then on professional/technical work. Managers' work is likely to be affected in at least four different ways. One obvious change will be that the first management jobs to be automated will be the very routine tasks. Laborious tasks of checking through figures and searching for data that relate to the control of people or functions under the manager's direction are those most likely to be the first to be put on the manager's personal computer or provided by the automated office system to which he or she has access. The ratio of routine to non-routine work will therefore be reduced. A second effect is likely to be a reduction in the number of people under the control of any one manager and an increase in the amount of capital equipment. This will to be most marked in offices. At the moment most office work is extremely labour intensive. In 1980 the average capital investment to support a typical UK office worker was approximately £500, as compared with £5,000 for their blue collar counterpart (TUC, 1979). The introduction of new technology in the office is already increasing the ratio of capital to labour cost, and it seems likely that this trend will continue. Similar trends have been observable on the shop floor for many years, and the issue of new technology having the effect of displacing large numbers of workers, which we discussed in Chapter 2, means that the trend is almost certainly acce-lerating. This means that, in both offices and factories, man-agers are almost certainly going to be responsible for more capital equipment. Given that there are limits to the amount of responsibility any one person can exercise, it is likely, other things being equal, that they may be expected to have re-sponsibility for fewer bodies. Spans of control are likely to be reduced.

A third factor changing the nature of managerial work, and associated with the introduction of more capital equipment, is that the qualifications of those being supervised by the average manager are likely to increase. With the more routine work automated and with more capital equipment to be operated, the number of unskilled jobs is likely to fall. This overall trend in industrialised countries can be established from censuses of occupations over the years, which show a steady rise in the number of qualified workers in the workforce. Despite ar-guments about de-skilling which were discussed in Chapter 4, we suggest that this trend is likely to continue and accelerate, and therefore result in managers having a more qualified work-force to manage. A fourth factor, associated with this last factor, is that the work done by these more highly qualified sub-ordinates will be more complex than in the past.

In all, the change in managerial work brought about by new technology can be characterised as being from a situation where the manager has to organise large numbers of relatively un-qualified people carrying out relatively simple tasks to one where the manager is responsible for a a complex system com-prising capital equipment and relatively well qualified people, where day-to-day operations are routinely carried out by the automated system. The managerial task will be to ensure the smooth running of the system, to exercise initiative in at-tempting to improve its operation, and to handle relationships with other units in the enterprise. An archetypal example of this kind of change is provided by the production manager of an electronics factory interviewed by the author. Fifteen years ago, he remarked, he was in charge of several hundred women making electronic valves. His main concern was to ensure high productivity from each operator and his obsessions were with turnover, absenteeism, and the bonus system. The product technology changed and the women then made transistors. This involved the use of some capital equipment and his concerns then had to include the maintenance of this. Now the transistors are produced by almost wholly automated machinery. There is only a handful of production operators within the production manager's responsibility, but he now also has a large group of technical people responsible for the operation of the automated equipment. Turnover and absenteeism are no longer the big issues, and the bonus system is virtually irrelevant to the productivity of the machines. His task is now that of managing this complex production process and its relations with other departments in the factory, such as materials supply, sales, and the finance department.

This inevitably makes new demands on the manager and the organisation. The manager now has to operate with a different style of supervision. He or she is likely to shift from a strategy of direct control to a strategy of responsible autonomy.[1] The better-qualified subordinates doing more complex tasks which require more judgement, and whose productivity is not tied to the production line, are likely to require more autonomy; there should be higher trust between manager and subordinate; and a participative management style is likely to be more effective. A different organisation culture is likely to be more appropriate: one which is less formal, less rule bound, more flexible, and task centred rather than role centred is likely to be more effective in the new circumstances.

Changes in professional/technical work

We note here three possible effects of new technology on technical work. One may be that it has a deskilling effect on technical work. A second is that it may lead to different technical groups becoming more closely integrated. Thirdly, it may lead to a new division of labour among technical staff.

[1] The terms are those used by Friedman (1977), following Trist *et al.* (1963), to describe, on the one hand, Tayloristic methods of control and, on the other, the Tavistock Institute socio-technical systems approach.

The deskilling effect we explored in Chapter 4 and no more needs to be said here except to note that, even if new technology leads to an increase in the overall skill level of the population, there will certainly be particular skilled groups whose jobs become de-skilled or disappear with the advent of new technology.

The closer integration of different technical groups may come about for a number of reasons. It may be that the new technology allows the enterprise to do more complex things. For example, an engineering company may be able to design and manufacture a more complex piece of machinery because CAD and CNC machine tools can now draw and then machine much more intricate shapes. This means, however, that those responsible for the machining and those responsible for the product design must be in closer contact than before in order to establish just what the technologies in the drawing office and on the shop floor are capable of. Moreover, in many applications of new technology, different departments in the one enterprise share a common data base. With several parties able to have access to, and manipulate, the same data at the same time, these groups need to operate in a more integrated fashion than under a system where files were passed manually around and work proceeded in a sequential rather than interactive manner.

Thirdly, the new technology may create a new division of labour among technical staff. New jobs are likely to be created. With the introduction of CNC machine tools, for example, the job of part programmer (the person writing the program which controls the machining operations) was invented. In drawing offices the division of labour between the designer and the draughtsman may be becoming blurred. In other instances, the introduction of new technology may well create, in theory at least, a number of possibilities for reorganising the way work is done. An important question both for management and for technical staff themselves is what impact has the new technology for traditional definitions of what constitutes the domain of tasks for each occupational group, and what opportunity does the new technology offer to management to create jobs which are filled by others besides those in the occupations traditional to that area. If such opportunities do exist, then will these new jobs lead to the creation of new occupations, with the job-holders developing a specific occupational identity and an established pattern of training being set up, or will management retain the prerogative to appoint whoever it chooses to these jobs and to supply whatever training it deems appropriate in an *ad hoc* way?

Changes in clerical and secretarial work

Perhaps the most obvious change to clerical and secretarial work which new technology is bringing is the very much higher productivity that it allows. All the predictions are that there will be a substantial reduction in numbers employed in these categories. Concomitant with this is the evidence beginning to emerge of a move towards pooling secretarial and typing work. If typing productivity is trebled, as some commentators suggest, then in situations where a manager has his or her own secretary, and that secretary spends most of his or her time typing (though this is not always a correct assumption), if there is not an equivalent increase in the amount of work to be typed then office managers will be tempted to pool typing resources in order to reap the benefits of the enhanced productivity. Thus one effect of new technology on secretaries and typists may be a fall in the number of job opportunities and a shift in their location of work from the boss's outer office to the typing pool.

All this assumes that the basic tasks performed by secretaries will not be altered as new technology is introduced. But what the new technology will do is to place VDUs in front of secretaries and typists. These terminals will, in many cases, have powerful information processing capabilities or be connected to central computers with even more extensive applications. It is possible that those who will be closest to these work stations (often the secretaries) will have the opportunity of performing more complex tasks associated with them. There is the potential, in other words, for renegotiating the division of labour between the manager or technical expert on the one hand, and the clerk, secretary, or typist on the other. An automated office system might be installed in the finance and costing department of a manufacturing enterprise, for example. Under the old system a secretary would type up tables of cost accounting data from handwritten tables prepared by the cost accountants. With the new technology the data would be tabulated on the screen of the automated system. If the data being tabulated is generated within the same system then it is conceivable that the secretary could call up, collate, and tabulate that data, following a set of rules which would probably be set out by the cost accountant, but without the involvement of the accountant. Whether this is likely will depend on a number of factors we have not yet discussed.

Changes in the nature of managerial control

It does of course depend on the way managements choose to use the new technology, but, if the analysis above is at all correct and the result of putting in new technology is to automate routine tasks, use a higher ratio of capital to labour, and to employ people with higher qualifications doing more complex jobs, then Taylorism as a form of management control will be dead. It will no longer be possible to exercise control based on task fragmentation, tight specification of procedures, and insistence that subordinates closely follow the rules. Instead, control will have to be exercised by developing higher levels of commitment to the enterprise and higher levels of trust between managers and their subordinates, and by management specifying goals and performance targets for complex sets of tasks. A priority management task of the future may well become that of defining tasks, work groups, measurable goals, and performance data capture systems to maintain effective control under the new conditions. Management is also likely to make more extensive use of internal labour markets as a control device.

Effects of new technology on size

The task of the enterprise is one of the factors normally thought to influence organisation structure. A second factor is that of the size of the enterprise. We turn now to a discussion of some possible effects of new technology on the scale of operation of work units, a discussion that is of twofold importance, for not only may size be an important influence on the organisation structure, but it is also an issue of interest in its own right.

To discuss the relationship between size and organisation structure we need first to make clear a distinction made by the economists who measure such things—namely, the difference between an enterprise and an establishment. In their terms the *enterprise* is the entire undertaking, including the Head Office and all subsidiary companies, whereas the *establishment* refers only to individual plants (places of work) within such enterprises. That this distinction is important is illustrated by the fact that there has been in the post-war period in Britain, as elsewhere in Europe and in the USA, a substantial decline in the number of people working in very large establishments. There has not, however, been an equivalent fall in the number of people working in very large enterprises. We should not, therefore, infer that the giant multinational corporation is becoming less important and powerful as an institution.

The decline in the number of people in large establishments has come about for a number of reasons. Some basic manufacturing industries, such as textiles, steel-making, shipbuilding, and car manufacture, have declined in the UK as production has gone overseas. Also, labour productivity in these industries has increased faster than output and there has been a shift in employment from these 'smoke stack' industries into other more recent manufacturing industries and, more especially, into the service sector. In general, these newer industries, though often owned by giant corporations, operate from smaller establishments. Another reason for the decline in large establishments has been changes in the composition of the workforce. There has been a steady increase in the proportion of more qualified/skilled people at work. Whereas in the past in many industries it was necessary, in order to achieve various economies of scale, to have large numbers of workers performing relatively routine operations using muscle-power on large pieces of equipment/products, now, with much of this kind of work done by machine and a higher percentage of people in staff functions, it may be easier to disperse these various functions in separate establishments. Very many of these smaller establishments—for example, hotels and other service-sector places of employment—are owned by large conglomerate companies, and thus their share of employment and their economic importance and power have not been reduced.

Predictions about the future size of enterprises and establishments ought to proceed from an analysis of the determinants of their present size—an exercise that is of some interest because of recent theories attempting to account for size differences. We review four of these theories now.

The first is the standard economic analysis in terms of economies of scale. This argues that, for any particular industrial operation, there is a certain size below which it is relatively uneconomic to operate. Examples of scale economies, to use the motor industry to illustrate the point, are the costs involved in designing a new model and the tooling required to make it. In order that these total fixed costs should be profitably recouped, several hundred thousand cars of that particular model need to be sold per year. However, it has been the case for a long time that many companies are much bigger than is necessary to achieve full economies of scale. Moreover, even in the case of car production just cited as an example, economies of scale do not explain why some components should be made in-house while others are bought in. Further explanations for the existence of large-scale enterprise are necessary.

One such is that put forward in 1967 by J. K. Galbraith, who argued that it was modern technology which led to the emergence of giant corporations. Such technology—he had in mind such twentieth-century developments as oil refineries, car production, and other complex mechanical and electrical/electronic devices—requires long lead times and much capital investment to develop, is inflexible once it has been developed (an oil refinery or chemical plant can only produce the product for which it has been designed, for example), and requires a complex organisation staffed by highly qualified and specialised personnel. Although such technologies usually need large plants for normal economies of scale, Galbraith's point is that corporations operating these technologies are, as we have noted, usually far bigger than the usual economies of scale argument would predict. He claims that this is because the inflexibilities of these high technologies, to which we have just drawn attention, make it impossible for firms to operate in a free market context. The unpredictability and volatility of the demand and price signalled in the free market is such that no firm would invest in these technologies under these conditions.

Markets therefore have to be fixed, by operating cartels or, more usually, by horizontal or vertical mergers, so that demand for the product is assured before the capital investment with the long lead time is made. It is the mergers which create the giant enterprises—mergers made not for reasons of monopoly profit, according to Galbraith, but in order to reduce risk-taking to an acceptable level when operating with high technology.

c) Control
with
Size

A third explanation for the large size of many modern business corporations comes from business historian Alfred Chandler. The title of his definitive study of American business history, *The Visible Hand* (Chandler 1977), expresses the germ of his thesis,—namely that, when the volume of transactions between two stages of a production process reaches a certain point, the visible hand of hierarchical organisation becomes more efficient than the invisible hand of the market place. In other words, when the volume of trade is rather small, it is more economic for manufacturers of component parts to sell to assemblers, who in turn sell the assembled product to the wholesaler. As the volume of trade increases, as it did very rapidly in the USA as the railroads opened up the country, there may come a point when it becomes more economic for the component producer to merge with the assembler and for them to set up their own distribution network. They will now co-ordinate transactions between the various parts of the merged enterprise using the newly installed management hierarchy and not the market place. This leads to increased size of enterprise for two reasons. Firstly, the vertically integrated enterprise is of course larger than each of the independent companies now merged. Secondly, there are economies of scale in the costs of administration. As co-ordination by hierarchy, for a large volume of transactions, is cheaper than co-ordination by the market, this gives a comparative advantage to the larger merged firm, which is likely therefore to increase in size still further.

The fourth explanation for the growth of large enterprises is that developed by the economist, Oliver Williamson, though he bases his theory partly on the behavioural approach to organisation theory developed by Herbert Simon (1961). This,

4) Transaction though it has its shortcomings (see, for example, Francis *et al.* 1983), is the most powerful and general of the four approaches,

and encapsulates most aspects of both Galbraith's and Chandler's approaches. Williamson, following an earlier economist Coase (1937), suggests that the most fruitful point of departure for explaining the existence of firms and other hierarchically organised institutions is the study of transactions, and the costs associated with them. Markets and hierarchies are alternative instruments for completing a related set of transactions, and whether a set of transactions ought to be executed across markets or within a firm depends on the relative efficiency of each mode. So far this parallels Chandler, but, whereas Chandler focuses only on the volume of transactions as the determinant of relative efficiency, Williamson introduces wider considerations. In particular, he draws attention to a general attribute of human behaviour, that of opportunism. This he defines as 'a lack of candour or honesty in transactions, including self-interest seeking with guile' (Williamson 1975, 9), and he assumes that most people most of the time will act opportunistically in the conduct of a transaction, given the opportunity. The opportunity for this behaviour is afforded by two sets of conditions. The one is that there is a fair degree of uncertainty or complexity surrounding the transaction—so much so that the limited capacity of the human brain cannot consider all eventualities (the 'bounded rationality' condition: Williamson 1975, 24). The other is that there are not large numbers of potential buyers or sellers associated with the transaction. Only when the so-called 'small-numbers' condition obtains (1975, 27) can one party to the transaction act opportunistically. If I buy apples from the market every week, and there are a number of greengrocery stalls, no one greengrocer will gain much of a long-term advantage over me by consistently exaggerating the quality of his fruit. Such opportunistic behaviour will only be worth while, to the greengrocer, if engaged in at the expense of the foreign tourist who is unlikely to visit Oxford market more than the once.

From these conditions, Williamson goes on to argue that, when the transaction between two stages in a production chain involves only small numbers and has a high level of complexity or uncertainty, the temptation to behave opportunistically will be high enough to make the parties to the transaction prefer it to be managed via a hierarchy rather than through the market place. If the two companies merge, and handle the transaction between the two stages of the production process by managerial hierarchy, then the merged entity has nothing to gain by either party behaving opportunistically in the conduct of the transaction.

The question is then why there has been a secular trend towards hierarchy and away from market. Both Galbraith's 'new industrial state' and Chandler's 'visible hand' can be reinterpreted within the Williamson framework. The various features of modern technology which Galbraith identifies as leading to inflexibility create, in Williamson's terms, the small-numbers condition. A firm which has invested many years of research and a great deal of capital in a technological development that is specific to only one product—say the engine for an automobile—and for whom there are only a limited number of buyers of that product, is very vulnerable to opportunistic behaviour on the part of the buyer—the car assembler. Similarly, the car assembler, perhaps having given the contract to develop the engine in a market where there were a number of bidders, is now faced with a single supplier of the engine and is vulnerable to opportunism on the part of the engine supplier. In these circumstances, Williamson would predict that neither party would be happy to take out a contract: either both would prefer to merge, or the car assembler would decide to develop and produce the engine in-house. It is thus not so much the general unpredictability of markets, but the specific threat of opportunism given the particular circumstances, which has led to the growth of the firm.

Similarly Chandler's general thesis that it is the volume of transactions which leads to hierarchy replacing market can be substantially modified. It is not the volume of transactions *per se*, but that in the instances Chandler cites there are also developments in the technology of process or of product, or in the market, which lead to the small-numbers and complexity/ uncertainty conditions. Where these conditions are still absent, mergers have not taken place and large volumes of transactions have taken place through the market place. An obvious example is the lack of vertical integration in food manufacture and distribution, where the few giant supermarket chains in Britain, and the USA, have not integrated backwards into manufacture, despite the huge volume of transactions each conducts with the suppliers. It is only where Williamsonian features obtain or where there are 'power' advantages to one party to the transaction that vertical integration takes place.

If one accepts Williamson's transactions-cost approach, then the question to be asked is what effect new technology is likely to have on bounded rationality, and the small-numbers condition. Prima facie, the bounds on rationality are likely to be reduced as electronic data bases and automated information processing systems are introduced. More information can be handled per unit cost, and therefore decision-makers may be expected to collect and process more information than before for any given decision. Also, one important instance of the operation of the small-numbers condition is that one party to a transaction may have to make transaction-specific investments. In many instances the transaction specificity of new investments may be reduced with new technology. In engineering, for example, the increasing use of computerised design and draughting equipment and of CNC machine tools gives a flexibility to manufacturers so that new or, more usually, improved or adapted designs can be produced and machined with lower set-up costs. Drawings can be revised on a CAD system much more quickly, and CNC machine tools require fewer jigs and other special tooling. Across a range of industries, economic batch sizes are likely to fall as set-up costs (the transaction-specific investments) reduce. The general argument is, therefore, if the bounds to rationality increase, and the small-numbers condition becomes less binding, then information impactedness will occur less frequently and there will be fewer occasions when hierarchy offers an advantage over market for the conduct of a transaction. We would therefore expect to see, if the transactions cost argument has any explanatory power, markets begin to replace hierarchy again. We would expect this to occur both through existing firms uncoupling various parts of their operations and reverting to transacting through the market place rather than via internal management hierarchy, and through new entrants to the market setting up in competition with parts of existing firms. If existing firms uncouple then, in many cases, this might occur by firms moving over to a sub-contract arrangement for the supply of various goods and services while still retaining the core activities. In other cases, one might expect to find companies hiving off individual business units, perhaps through management buy-outs, or to other more conglomerately organised companies. There is some evidence of this already happening though there are obviously a number of other reasons which also help account for this currently widespread phenomenon.

Telecommuting

Telecommuting clearly has implications for the size of establishments, and the extent to which telecommuting will occur in the future will be influenced by new technology for both technical and transactions cost reasons.

Telecommuting is already emerging as a trend, and we present some of the evidence in Chapter 8. One obvious feature of the new technology which is making this possible is the changing nature of the work people are doing and the available technology to do it. Computer programming itself as a task is an obvious occupation to perform at home. The only equipment necessary to do the job is a computer terminal. The task of computer programming is complex but fairly predictable so that programmers can be set realistic deadlines to complete programs but do not need close supervision while they are working. It is not surprising that this particular task should be the first to give rise to telecommuting on a large scale. In fact it has done so in a way which may be archetypal for new forms of work organisation. There is now an agency in the UK, F. International, which offers client companies a computer programming and consultancy service by acting as broker between the client and a pool of over 700 free-lancers working from home. As compared to a company having its computer programs written by its own in-house full-time information systems department staff, this is market replacing hierarchy on a grand scale.

But the new technology not only provides the technical means for telecommuting—the cables linking home and office and the fact that so much work today is to do with manipulating information rather than things—it also provides the means of conducting transactions at a distance. We have already noted in our discussion of how companies handle transactions between stages of the production process how the new technology is shifting the comparative advantage away from hierarchy and towards market by reducing the constraints of bounded rationality and transaction-specific investments. In a precisely similar way the new technology may be shifting comparative advantage from the internal labour market to the specific short-term contract between an individual worker and employer. If this is the case then it is likely that some firms, and some individuals, will prefer to work from home on a free-lance basis rather than enter into a long-term employment relationship. The important question is whether the general case will be that firms and individuals will share the same preference. We showed in Chapter 4 that in the early days of the Industrial Revolution manufacturers preferred to replace the free-lance contract and get people into the factories on an employment basis, and that this went strongly against the preferences of the workers themselves. It is tempting to suggest that today the position is reversed, with many workers preferring the security of an employment relationship, buttressed as it is with a certain amount of employment protection legislation and backed up by a degree of trade union strength, whereas firms would prefer to revert to a contracting-out arrangement.

Effects of new technology on organisational shape, flexibility, and complexity

We have already mentioned, in the previous chapter, the tradition within organisation theory which suggests that an enterprise's organisation should be matched to the task done by that enterprise, and the technology it uses. We now sketch out some possible connections between new technology and new forms of organisation based on the implications of the work of four leading organisation theorists.

The earliest attempt to link technology and form or organisation was that made by two researchers, Tom Burns and G. M. Stalker, who more or less stumbled across the finding in the course of a sociological investigation of the social structures of a rayon mill and an engineering concern. At the time of the research, 1961, it was universally believed (at least among teachers of management, if not the practitioners) that there was one best way to manage—a system encapsulated in the principles of classical management theory. Though Burns and Stalker were not themselves teachers of management, they were familiar with the precepts of classical management theory and were unsurprised to discover that the rayon mill under investigation was organised in this manner. Within the engineering concern, in contrast, they found that positions and functions in the management hierarchy were ill defined, and that this was so because of the deliberate policy of the head of the concern. Though this led to considerable feelings of insecurity and stress among individuals in the organisation, and to people spending much time and energy in internal politics, the firm was a commercial and technical success. The researchers wondered whether the firm's success would have been even greater if they had sorted out the organisation structure and reduced the level of stress and anxiety, or whether, on the contrary, the insecurity, stress, and anxiety might not be the mainspring rather than the grit in the management system, and a contributory factor in the success.

These questions became the focus of a series of further investigations into nearly twenty other firms, and led Burns and Stalker to the conclusion that there seemed to be two divergent systems of management practice and forms of organisation structure, though in no case did either system seem to be adopted as the result of an open and conscious management policy decision. They labelled one of these systems 'mechanistic' and the other 'organic' and suggested that the former was more appropriate to circumstances characterised by stability, whereas the organic form was more suited to situations where there was rapid change, for example in the markets being served or in the technology being used in the product or production process.

Burns and Stalker describe the differences between the two systems thus:

In terms of 'ideal types' their principal characteristics are briefly these:

In mechanistic systems the problems and tasks facing the concern as a whole are broken down into specialisms. Each individual pursues his task as something distinct from the real tasks of the concern as a whole, as if it were the subject of a sub-contract. 'Somebody at the top' is responsible for seeing to its relevance. The technical methods, duties, and powers attached to each functional role are precisely defined. Interaction within management tends to be vertical, i.e. between superior and subordinate. Operations and working behaviour are governed by instructions and decisions issued by superiors. This command hierarchy is maintained by the implicit assumption that all knowledge about the situation of the firm and its tasks is, or should be, available only to the head of the firm. Management, often visualized as the complex hierarchy familiar in the organization charts, operates a simple control system, with information flowing up through a succession of filters, and decisions and instructions flowing downwards through a succession of amplifiers.

Organic systems are adapted to unstable conditions, when problems and requirements for action arise which cannot be broken down and distributed among specialist roles within a clearly defined hierarchy. Individuals have to perform their special tasks in the light of their knowledge of the tasks of the firm as a whole. Jobs lose much of their formal definition in terms of methods, duties, and powers, which have to be redefined continually by interaction with others participating in a task. Interaction runs laterally as much as vertically. Communication between people of different ranks tends to resemble lateral consultation rather than vertical command. Omniscience can no longer be imputed to the head of the concern. (Burns and Stalker 1966, 2nd edn., 5–6.)

The implication of all this for a discussion about the effects of new technology is, of course, that it is going to be much more appropriate for enterprises to operate with an organic structure if they make extensive use of this technology. If 'the management and technical tasks are as characterised at the beginning of this chapter, if microelectronics-based technology takes over much of the routine work, if the speed of change of the technology maintains its current pace, and if this continues to cause rapid changes in product markets, then, according to Burns and Stalker's evidence, firms should operate with organic management systems. Those enterprises most affected will be those which, up to the present, have found it most appropriate to operate with a mechanistic system. Burns and Stalker's organically managed engineering company was in the electronics industry. Many engineering companies, especially those in the mechanical and heavy electrical engineering sectors, will have been operating for very many years with highly mechanistic structures. These may have been entirely suitable for firms in mature market segments where the rate of change in products, processes, and markets has been very slow. If, as seems the case, the impact of the new technology has been to introduce radically new production systems (for example, CAD/CAM and flexible manufacturing systems (FMS)), and these have also had implications for markets and, to some extent products, and these changes are likely to continue happening, then a switch to a more organic system of management may be very necessary for survival. For firms which have been used to the mechanistic form for decades, and staffed by individuals who have had a lifetime's experience of operating in this particular way, such changes will be extremely hard. Those companies which can manage the change effectively and swiftly are likely to gain a significant competitive advantage.

We have already mentioned, in Chapter 3, Joan Woodward's classic research into the management practices of over 100 firms in South-East Essex. This was carried out at the same time as the Burns and Stalker research, though neither team was aware of the other. Like Burns and Stalker, Woodward's discovery of the link between technology and organisation structure was serendipitous. Following the initial findings of this research, Joan Woodward moved on to Imperial College, London, where the team she built up around her continued work into the link between technology and organisation structure. It was this group which developed the notion, which we have referred to already in Chapters 3 and 6, that the management control system is the crucial intervening variable between an enterprise's technology and its organisation structure (Woodward 1970). These control systems vary along two dimensions, she suggested: they may be exercised through an impersonal control system (for example, written-down rules and procedures or electronic control systems) or through communication down a hierarchical chain of authority; and the control system may be fragmented (in the sense that there are a variety of objectives and associated control systems within the organisation which are not integrated) or unified. Where the control system is located along these two dimensions in an enterprise will have a

very significant effect on the nature and quality of the relationships between the various organisational members.
Though management have some degree of choice over what
type of control system they operate, the task and technology of
the enterprise also have some bearing on the matter. Thus, it is
argued, the introduction of new technology may have a profound influence on organisation structure.

It is tempting to predict that the introduction of new technology will shift organisations from the use of impersonal, fragmented systems for controlling *people*, that impersonal systems
will become unified through the application of electronics-
based information technology and be applied only to machines,
and that people will be co-ordinated on a personal basis with
a concomitant improvement in the quality of working life.
However, the choice of control system is a managerial one.
Managers will not necessarily know what system is the most
efficient, and they may be guided by the kind of preferences and
concerns we discussed in Chapter 6. A later, more sophisticated,
attempt to understand the relationship between technology
and organisation is that of the Yale sociologist Charles Perrow
(1970). In fact Perrow's analysis is in terms of the tasks an
organisation does rather than the technology it operates, but
there is, of course, a close link between the two.

Perrow suggests a fourfold typology of tasks based upon two
independent dimensions, the degree of variability and the degree of uncertainty in search procedures, and he links each type
of task with a set of organisational characteristics. For example,
tasks which have a high degree of variability and where the
level of uncertainty is so high that the search procedure for
solutions cannot be codified are typical of those performed by
professional occupations. Perrow's argument is that tasks at
this extreme on each of his two dimensions are best managed
by forms of organisation typical of professions—that is, a high
degree of collegiality, co-ordination by mutual adjustment, and
virtually no hierarchy. Tasks at the other extreme are charac-
teristically organised by the most mechanistic forms of hierarchical bureaucratic organisation.

The implication of the Perrovian analysis for our discussion
about new technology would appear to be that if new technology gives management more understanding about the na-

ture of the task being done by the enterprise then more bureaucratic structures can be installed. This would, however, be a simplistic application of the analysis. It is more likely that new technology will allow the automation of the more certain elements of the task, so that those left are the ones with higher levels of variability and uncertainty. If this is the case then new technology is likely to lead to less bureaucratic structures. Perhaps the safest prediction to make is that new technology may allow management a degree of strategic choice—a possibility we explore at greater length below.

Our fourth organisational theorist is Jay Galbraith of Harvard. His argument is that organisations exist to process information. Defining uncertainty as the gap between the information an organisation needs to complete the tasks in hand and that which it already has, Galbraith (1977) argues that if the level of uncertainty facing an organisation rises then it should change its form of organisation in specified ways. With low levels of uncertainty, it is enough to rely on standardisation, the use of rules and procedures, and co-ordination via the managerial hierarchy. As uncertainty increases, a need arises for planning by the organisation. Beyond that, as uncertainty increases still further, the enterprise is faced with a choice. It can make moves to contain the amount of information processing required by introducing the use of slack resources or setting up self-contained task units (that is, moving from a functional to a product-based form of organisation), or it can make changes to its organisation in order to enhance its information processing capabilities. It does the latter, suggests Galbraith, by strengthening its vertical managerial hierarchy or by setting up formal horizontal communication channels across the enterprise. Examples of this second strategy are the use of liaison positions, task forces, or project teams, or, ultimately, the institution of a matrix organisation structure.

Galbraith sets out the same range of possibilities as Perrow, and there appears to be strategic choice to be made by management about the use they make of new technology. On the one hand, new technology use may increase the level of uncertainty which the organisation faces because, for reasons we explore later, it increases the amount of information which the organisation needs to handle. *Ceteris paribus*, the organisation is then likely to become more complex to cope with the extra uncertainty. Alternatively, because the organisation may use the new technology to process information within the enterprise, the level of uncertainty may fall, in which case the form of organisation may simplify. A third possibility is that an enterprise may be prepared to operate with a level of complexity in its organisational form that is determined by such factors as the ability of organisational members to cope with that level of organisational complexity, the cost of running that type of structure compared to the cost structure associated with the product market in which it operates, and the knowledge of senior management about organisational possibilities. If this is the case, and if changes in new technology shift upwards the ability of the firm to cope with a given level of uncertainty, then the impact of new technology will be that the enterprise attempts to deal with more information relating to its environment—that is, it will attempt to do more complex things, more cleverly, and more efficiently.

Conclusions

The major implication of the above analysis is, on the face of it, that, although hypotheses might be generated on the basis of agency theory, TCA, or contingency theory (namely, that new technology will lead to more telecommuting, more freelancing, and more complex high-trust forms of organisational arrangements), these differ from the hypotheses that would stem from an analysis of the exercise of power within and around enterprises. While some applications of new technology may lead to the former kind of hypotheses, because they allow control to be exercised at a distance through discrete information gathering, or because they allow tasks to be more highly programmed, other applications may lead in the opposite direction.

To conduct the analysis in a specific case would involve at least the following three steps, not necessarily in the order given. One would be to conduct the conventional analysis as set out in the early part of this Chapter, looking at the implications for the particular application of new technology for metering, opportunism, and reduction of uncertainty. The next step would be to look at the implications for the four strategies detailed in Chapter 6 in the section entitled 'Preferences of capital and labour within each control type'. To what extent might the new technology be used to break union solidarity? Management may well try to use the introduction of new technology to oust old occupational groups and introduce new categories of work over which they, rather than occupational members, can maintain control. The attempt to do this may mean that a form of work organisation is adopted which would not be the one predicted by a conventional organisational analysis. To what extent will the technology be used to fragment tasks so that a variety of different-priced labour can be used? To what extent will the technology be used to break down labour opportunism so that the effort–reward bargain can be renegotiated? To what extent can it be used to 'drive' workers harder in the way Marglin describes in his history of the rise of factory production? At least equally important, and arguably more important, is to pose the converse of these questions. To what extent may new technologies *not* be taken up in the way a rationalistic analysis would predict because they would not allow management to continue controlling labour in the direct ways implied by the questions above? The issue of telecommuting is a case in point. Many firms acknowledge that they already have jobs which could be done from home on a telecommuting basis but they are not operating in this way. Arguably some of the control strategies we have identified could not be exercised so effectively on home-workers. Many new technology applications may give workers greater possibilities for acting opportunistically against management, and if management realise this possibility this will affect how the technology is taken up.

The third part of the analysis, though, is to examine the question of how important the issue of control over labour is in specific instances. It may now be the case in many potential new technology applications that high capital utilisation, maintaining presence in a particular product-market niche, or keeping up a high rate of innovation are much more important objectives for management in terms of maintaining profitability than getting the last ounce of routine effort out of each worker. In such cases, management may either be relatively unconcerned about the implications of new technology for control or be concerned to balance control issues against motivational issues. The way in which the labour market is operating at the particular point of time (in terms of workers' propensity to quit) may be an important factor in the analysis.

Our conclusion, therefore, is that an exploration of the two-way relationship between new technology and organisational form requires case-specific analysis, taking into account all the factors and issues suggested here. It is an analysis that takes us beyond the technology itself and the enterprise contemplating its use, and involves a consideration of capital, labour, and product markets, and trade union power and policies.

8

New Technology: The Challenge to the Unions

The nature of the challenge

Many of the changes at work brought about by new technology that we have described so far in this book have severe implications for trade unions. In the first part of this chapter we spell out what some of these are and in the second part we discuss how unions are responding, and what other possible responses could be made.

The most obvious challenge to organized labour is the labour-displacing tendency of new technology. Even if they take the optimistic view that in the long term economic forces will restore the economy to an equilibrium, high, level of employment, unions will be involved at local level in the introduction of new technology which will reduce labour requirements in specific plants at specific points in time. If all labour-reducing technical changes are opposed then unions are open to the criticism that, by opposing technical change, they will make industry uncompetitive and so cause the loss of even more jobs through firms going out of business.

A second challenge to the unions is the change in plant size itself that results from the introduction of new technology. The research evidence, and common sense, indicates that it is easier for a union to develop strong organization in large plants, and union density (the proportion of workers who are members of a trade union) is higher in larger plants. Full-time officials can service one large plant more effectively than several small ones. In large plants it is more efficient for both management and unions to develop rules and procedures to govern pay bargaining, grievance procedures, etc. than to operate on an individualistic basis with management treating each worker as a specific case.

Thirdly, there is growing evidence of the growth of a diverse range of employment contracts. There appears to be a reversion to some extent to the variety of ways of contracting with labour which existed in the last century. Hiring skilled and professional workers on a contract basis is one example. There is evidence, as we have shown, of considerable use of outworkers with new technology, and every reason to suppose that there will be some increase in telecommuting. For the reasons we have already discussed—principally the greater information handling capabilities of the new technology which allow contracts and control to be specified and exercised more tightly—these alternative forms of contract may proliferate. The European Foundation for the Improvement of Living and Working Conditions (1984) has identified the following list of variants to the full-time 35–40 hour week based in a work-place or office within the employing organisation:

(*a*)   Contract work.
(*b*)   Outwork.
(*c*)   Part-time work.
(*d*)   Temporary work.
(*e*)   Permanent weekend work.
(*f*)   Twelve-hour shiftwork.
(*g*)   Full-time work extending over three to four days.
(*h*)   'On-call' work where the worker awaits the call at his or her home.
(*i*)   Isolated work.

Some of these changes, such as outwork, shiftwork, and some contract work, may be attributed to characteristics of the new technology itself such as the capacity it brings to telecommute, or, in some instances, its high capital intensity making shiftwork an attractive financial arrangement for management and, possibly, for some workers under certain conditions. Others of these changes may be the result of the shift to the service sector which the high productivity of new technology in manufacturing is speeding up. Weekend work and seasonal work are characteristic of employment in the leisure and recreational sectors of the service industry. Servicing the new technology itself will be a growing industry and likely to lead to increased on-call and isolated work.

Associated with these changes in the form of work and different contractual arrangements may well be changes in the payment systems used. Such changes may also come about within the more conventional employment contracts. If, as is likely, levels of productivity for blue-collar workers become increasingly dictated by the machinery rather than the efforts of the individual workers then payments-by-results systems are likely to be replaced. Whether they are replaced by systems which rate the job or which rate the worker will be a decision that a union will want to become involved in, presumably favouring the former system, over which they can exercise more control and which does less to enhance an individualistic relationship between management and particular workers. White collar workers, and those telecommuting and/or working on a contract basis for example, may on the other hand be subjected to a shift in the opposite direction. At present, most white-collar workers are paid a salary based on rating the person's past performance in general terms and their future potential. The increased use of new technology in white-collar work is likely to have at least two effects of relevance here. In some instances it will enhance senior management's ability to measure the productivity of individual white-collar workers. In others it will lead to such workers doing more complete tasks and so the contribution that worker is making to that task will become more easily measurable. In either case the payment system can be shifted away from a generalised rating of the person in the direction of a more specific payment by results. In the case of contract work this is the whole basis of the payment. The challenge for unions here is threefold. In the case of blue-collar workers the shift away from payments by results to rating the person weakens the union's collective bargaining position. In the case of white-collar workers, for those workers in employment, if there is a shift away from rating of the person to payment by results then this is likely, other things being equal, to increase the propensity of such workers to join a white-collar union. The challenge to any union here is to orient its strategy and structure in such a way that it will be able to recruit such potential members. For those white-collar workers who are likely to become telecommuters or to work on a contract basis, union membership is a less obvious possibility. Telecommuting, or to use the more old-fashioned but more evocative term, home-working, puts the individual workers in a weak position *vis-à-vis* the employer as they are so isolated. They lack knowledge about the rate of pay and form of contract offered to others in a similar position, and, in the past anyway, have found it difficult to communicate with other home-workers to establish whether any feelings of exploitation are shared. Unions for the same reasons have found it difficult to recruit home-workers. It is possible that more middle-class home-workers, equipped with sophisticated information technology, may be less vulnerable, better informed, and more in touch with others in a similar position than traditional home-workers.

Working on a contract basis has not, in the past, precluded strong union involvement. As Stone's (1973) study of the early iron and steel industry in Britain shows, the skilled craftsmen in that industry worked on a sub-contract basis but operated a powerful union. Indeed, as we have seen, Stone argues that it was the power of the union to control the contract that led management to break up the system of contract working and replace it with direct employment and extensive use of an internal labour market. It does not follow directly, therefore, that increased use of contracting for specialised skills in firms using new technology is an anathema to union organisation. Whether those working on a contract basis will be tempted to organise themselves will depend to some extent on their ideology, but also on the way work is organised. If many contractors have similar skills and are doing similar jobs, and there is little or no restraint on the supply of people capable of doing the work, then these are conditions under which one might expect unions to be formed or joined. To the extent that employers and managers have some control over these factors, one would expect intelligent managers, mindful of their own interests, to minimise the extent to which these factors were apparent to the people to whom they gave contracts to reduce the likelihood of their becoming 'union-minded'.

This brings us on to the fifth major challenge to unions, which is that of changes in the skill composition of the workforce and changes in the occupational structure. We have already discussed at length the de-skilling strategies which some people have proposed that management have engaged in as they have introduced new technology. We have shown that the evidence is that, by and large, the simple version of the de-skilling hypothesis is not supported. A major reason for management not adopting de-skilling as a strategy to exercise tight control over labour is that this form of direct control is not particularly effective for many forms of new technology. What management needs is the co-operation of workers. 'Responsible autonomy' from workers is the alternative which Friedman (1977), following Trist and Bamforth (1951), has proposed. To gain this co-operation management may well not have de-skilled workers, but they do seem to be using other strategies which have considerable implications for unions. One of these is the increased use of internal labour markets, so that workers are now more concerned about their individual career up the hierarchy in the firm rather than in collective action with all those others in the same position as themselves in the firm. A second is the replacement of traditional occupations by new ones. Work associated with new technology may be restructured so that it is performed by those who, while still operating with a high level of skill, are no longer members of a traditional occupational group. Indeed the job they do may no longer have a specific occupational title. Their job is defined by the enterprise. Training may also be given by the enterprise. This represents a total loss of occupational control and a gain in control by the enterprise. This process can be clearly seen at work in the development of the occupation of software engineering. Results from a study conducted by Dick Holti (forthcoming) but not yet published show that software engineers have not developed any institutional control over their own occupations and that major decisions about forms of training and qualifications are in the hands of the employers. There is every reason to suppose that this is a strategy that management will want to pursue, and it is one that poses a very severe threat, particularly to occupationally based unions, but also to all unions, as the individual now stands in an individualistic relationship to the enterprise rather than as one of many who can best be represented collectively. How do unions meet this challenge?

The union response

We have already noted in Chapter 1 that the time that public awareness of the challenge of the chip began in Britain can be pinpointed almost to the day—that is, to the screening of the BBC documentary in April 1978. Not only did government move swiftly in taking steps to develop policy about micro-electronics technology, but so too did the trade union move-ment. Such rapid response may have been aided by the fact that at that time the trade union movement in Britain was heavily involved in tripartite discussions with government and industrial management about a wide range of economic and industrial issues. Whatever the reason, the trade unions very quickly began to develop policy statements about the social and economic impact of this new technology. In the period 1979–80 the TUC and many individual trade unions published policy documents about the new technology, and developed recommended procedures for handling its introduction.

The TUC published their report, *Employment and Technology*, in 1979, and *Labour Research* in the same year published a review article 'Microelectronics—the trade union response' which examined eighteen separate agreements or proposed agree-ments made between management and trade unions. The union policy documents produced by that time were, in addition to that of the TUC, from the National Union of Journalists (NUJ), the Technical and Supervisory Staff section of the Amal-gamated Union of Engineering Workers (AUEW(TASS)), the Association of Professional, Executive, Clerical, and Computer Staff (APEX), and the Association of Supervisory, Technical, and Managerial Staffs (ASTMS). These are all white-collar unions. Most blue-collar unions have had a long history of dealing with technical changes and already had policies and procedures for negotiating the introduction of new tech-nologies. They took the view that microelectronics technology was not sufficiently different for them to justify developing new policies.

What, then, is the content of the TUC and individual union policy documents?

Without exception they take a positive line about accepting the technology itself. The TUC's document *Employment and Technology* sets the tone:

by looking at historical data, we can see a fairly clear pattern of high-productivity growth being associated with high-output growth, good trading performance and, in consequence, steady or rising employment. It would be complacent, however, to argue from such figures that rapid technological change and its associated productivity growth will by itself be sufficient to set us on the virtuous circle of high output, low unit costs and high employment.(TUC 1979, 12.)

In other words, the TUC was accepting the value of investing in technologies which were likely to increase productivity (through labour saving), in the belief that this in the long term would be good for the economy and, in turn, for employment levels. However, it did not believe that the high-growth path was likely to come about by leaving adjustment of the economy simply to market forces. The TUC policy document was written at a time when the then Labour Government was committed to a tripartite planning approach to economic affairs. The TUC document therefore called for

involvement in economic, industrial and social policy measures across the different levels of the economy, including national, industry, company and plant levels . . . continued work on the Industrial Strategy (agreed with the Government); expansion of services; the promotion of expansion in world trade and of research and development effort. (TUC 1979, 56 f.)

No individual union had as its official policy the rejection of new technology on job-loss grounds. Even the craft print unions were prepared to accept the technology itself; their particular defence against job losses was to insist on specific manning arrangements. The National Graphical Association (NGA), for example, insisted on retaining control of front-end systems: that is, only NGA members should operate the initial key stroking of data into the microelectronics-based printing machinery, even though it would often be easier and more efficient for journalists (NUJ members) to type their articles directly into the system.

Within the context of a general acceptance of the principle of installing new technology, the TUC and the unions (with very few exceptions) had a rather limited set of concerns, based on their traditional bargaining issues of terms and conditions of work, defined rather narrowly. They were concerned to protect the jobs of existing members; to ensure that the pressure of work was not intensified; to protect, and if possible to enhance, pay levels and hours of work; and to ensure that health and safety at work were safeguarded. In addition, the TUC used the opportunity presented by bargaining over new technology to advance again the arguments for industrial democracy in such decision making.

A device attracting widespread support in the union movement was that of attempting to negotiate a new technology agreement (NTA) when new technology was introduced to the work-place. The TUC document set out a checklist for negotiators of NTAs. They suggested it should cover such items as:

(*a*) Ensuring the continuation of 'mutuality' (the principle that any change in terms or conditions at the work-place should not be introduced until it had been mutually agreed between management and the union representative).

(*b*) Establishing and maintaining joint management–union consultation machinery about the introduction of new technology.

(*c*) Establishing and maintaining collective bargaining machinery.

(*d*) Insistence on adequate provision of information, establishing of agreed plans on employment and output, and agreement on retraining and earnings protection for individuals whose jobs change, or who lose their old jobs, when new technology is introduced.

(*e*) Agreement on hours of work.

(*f*) Agreement about control over work. (But this item in the TUC document is concerned only with the ability which new technology gives management to increase the regulation and control of work. The TUC gives only four lines to this item and these do not relate to the issues of management control dealt with in Chapter 6 of this book.)

(*g*) Health and safety.

(*h*) Procedures for reviewing progress. Beyond the establishment of NTAs the TUC also set out a programme for trade union action more generally. One item, as we have already mentioned, is the establishment of a greater level of industrial democracy. They advocated the setting up of joint inter-union bodies at plant level (inter-union because of new technology blurring boundaries between jobs) which would obtain information about corporate

planning and projected technical developments. NTAs
should be pursued. It should be the objective of unions to
seek reductions in the working week, the working year, and
the total percentage of one's lifetime spent in work. There
should be more training, and more trade union education
about the likely effects of new technology, and about how
to deal with them.

APEX was the one union which went further than the others,
and the TUC, in its discussion about the effects of new technology on job content, job design, and skills. In a commendable
chapter in its policy document (APEX 1979, ch. 4) it explored
a number of the issues we have developed in this book. This
chapter began by warning that management may attempt to
use work-study practitioners, when new technology is introduced, to introduce multi-skilling and flexibility into the work
organisation in order to break down 'restrictive practices'
(APEX quotation marks) and demarcation. It also warned
against job enrichment as a substitute for better wages. The
warnings given, the document then advocated new forms of
job design in order to increase job satisfaction. It set out the
Tavistock Institute's list of factors relating to good job design
and argued that union representatives should bargain for these.
The list is similar to that set out by Hackman and Oldham
which we have discussed in Chapter 3. The APEX document
listed specific areas for union attention on job design and
recommended APEX representatives should resist any further
fragmentation and specialisation of jobs. They should be extremely critical of any proposed specialisation and centralisation of typists', clerk–typists', and secretaries' jobs which
would de-skill, routinise, lower the status of, or isolate the
affected staff. The social costs to employees and the likely lower
productivity for employers should be emphasised. APEX representation should propose alternative schemes for the redesign
of jobs to improve the quality of working life—for example,
semi-autonomous work groups. Job redesign is not to be used
as a means of cutting the total number of jobs or to increase
'flexibility' as understood by many employers. It is no substitute
for better wages and working conditions negotiated through
normal collective bargaining procedures (APEX 1979, 39-40).

APEX also laid out the stages along which new technology should be implemented. They advocated the setting up of a joint management–union steering committee at the initiation of the change. This should then carry out an assessment of the change, and submit a report. There should first be a trial implementation, a pilot study, which should be assessed before the final implementation is made, and this final implementation should be subject to continuous monitoring.

Many of these guidelines have now been incorporated into an impressive, glossy, and well-produced booklet called *Job Design and New Technology: APEX Guidelines* (APEX 1985) which has been drawn up using advice from the Work Research Unit[1] and from academics. Though other unions have become concerned about job design issues, APEX is the only union to have made such substantial progress in this area.

By June 1979 *Labour Research* was able to report the existence not only of the five union policy documents to which we have already referred (those of the TUC, NUJ, AUEW(TASS), APEX, and ASTMS), but also eighteen separate NTAs or proposed agreements. All had as their basis that there should be no unilateral introduction of new technology by management alone; that there should be full consultation on the basis of full information; and that installation should only proceed after agreement by the union of the terms of the installation. Two of the model NTAs, those of the Civil and Public Servants' Association (CPSA) and the Post Office, specify a cold storage period at the end of the trial stage with a reversion to the old method of working (in these particular cases, back to standard typewriters after experimenting with word processors) while the trial is assessed. Only some of the agreements contain a commitment to no redundancies as a result of implementing the new technology, and those that do contain a 'no compulsory redundancies' clause usually accept that there will be some natural wastage after the systems are up and working. Agreements in the public sector, though not in the private sector, often called for union involvement in planning increased pro-

---

[1] The Work Research Unit (WRU) was set up in 1974 by the Department of Employment subsequent to the formation of a Tripartite Steering Group on Job Satisfaction (TSG). The TSG draws its membership from the Confederation of British Industry (CBI), the TUC, and government, and continues to evaluate and assist with the development of the WRU programme.

duction and new products to compensate for employment lost by the new technology.

Though the NGA had fought successfully to maintain sole rights to input text (as we noted earlier) and at Rolls Royce the white-collar section of the Transport and General Workers' Union (TGWU(ACTSS)) had obtained an agreement that 'the operation of visual display units . . . will have ACTSS as the sole representing body', only one NTA has specific reference to the possibility of management using new technology to tighten management control. Even that reference only touches on the surface of the problem. It is within an agreement with Ford Motor Company and simply states that 'All information acquired specifically or incidentally by computer systems shall not be used for individual or collective work performance measures'.

None of the model agreements contain proposed reductions in hours to be worked, though most propose a general increase in wages when new technology is introduced.

By late 1983 *Labour Research* was reporting that:

whereas 1978–1980 were years of frantic trade union activity at national level in the production of policy statements, model agreements and booklets for members on the problems of negotiating new technology, since then there has been a relative silence as the movement has progressed from propaganda to the problems of implementation. Hopes and aspirations [among trade unions] of the late 70's have not been fulfilled. (*Labour Research* 1983, 296.)

Labour Research reviews developments which have taken place over those four years and concludes that 'the job loss threat and speed of technological change, to date, has been much more acute in the office than on the shop-floor' (1983, 296) and so it has been white-collar unions which have faced the problem of changing working environment, in an organisational way, often for the first time.

Some early trade union hopes of negotiating from a relatively strong position to achieve the objectives enshrined in their policy documents have not been realised. The CPSA, for example, were not able to achieve all they wanted in the NTA they finally signed with government. There was, for example, no guarantee of no redundancy. The Manpower Services Commission (MSC) attempted to implement a system for linking job centres in London through on-line VDUs to a central data base which could match jobs to job-seekers. The unions insisted that the initially proposed job loss of 10 per cent was too high, and demanded that the resources freed by the system be used instead to improve the standards of service generally in job-centres. A compromise agreement was eventually reached with the MSC, but it was not accepted by the Treasury on the grounds that the agreement did not provide for sufficient savings to make the expenditure on the technology worth while. The project was abandoned.

Some other white-collar unions have been more successful in getting their demands met. The National and Local Government Officers' Association (NALGO) has perhaps been the most successful, having signed over 100 NTAs, one of which (one of the few throughout the country to do so) has negotiated a reduction in working time. There are at least three reasons for NALGO's relative success. Many of its branches are huge and deal with a single employer. Many of the local authorities who employ its members will be politically more sympathetic to negotiating reasonable agreements than other employers. Thirdly, many of the chief officers who have responsibility for implementing the agreements are also members of NALGO.

The Banking, Insurance, and Finance Union (BIFU), by contrast, has only been able to negotiate one agreement with a no-job-loss guarantee and has had members sacked at one company for refusing to work with new equipment. The AUEW(TASS) has faced very rapid penetration of CAD, and management have often been able to introduce double day shifts in the face of union policy opposed to this. In a comment that could well serve as a general summary of the unions' weak position in the face of the introduction of new technology generally, *Labour Research* remarks that 'The carrot of 25 per cent pay premium for shift work and the stick of recession have made it difficult for the unions to resist the growth of shift work' (1983, 296).

For manual unions robots are the most dramatic manifestation of new technology, but their introduction has so far been slow enough to have not yet affected large numbers of workers or their work-places. The numerically more significant impact has come from the development of 'next generation' equipment, notably CNC machine tools. This has not led to the same kind of response from manual unions as white-collar unions have developed. 'In many cases management have felt that they can use their stronger bargaining power to push new technology through unilaterally.' (1983, 297.) Though *Labour Research* claims that the TGWU has found a number of medium-and large-scale employers who have used that strategy but have subsequently come unstuck over the question of pay structure and been forced back to consultation and negotiation with unions over the whole issue, I am doubtful that this experience is widespread and suspect that many firms have been quite successful in pushing through new technology use without any formal consultation at all. Research conducted by the author and colleagues in a number of engineering plants has revealed examples of this.

In contrast to the limited attention given to the issue of job design in the 1979-80 period, by 1983 *Labour Research* was reporting that not only APEX but also ASTMS and the National Union of Public Employees (NUPE) were increasingly addressing themselves to the question of job design, though in the last case this was in opposition to management who were using job redesign as a weapon in the privatisation issue.

Home-working, too, is beginning to receive more union attention in Britain. A report by Ursula Huws to the Equal Opportunities Commission[2] showed for the first time the emergence in Britain of the new kind of home-worker that the European Foundation had predicted. Out of a small sample of 78 new technology home-workers surveyed, Huws found that the large majority were highly skilled professionals. Only 10 per cent were clerical operators (that is, operating word processors) and 75 per cent were programmers, analysts, project managers, or consultants. There have been a number of recent reports (*Labour Research*, Nov. 1983; the *Guardian*, 15 Apr. 1985) of a company called F. International, the vast majority of whose 700 staff are free-lance, working from home, and 90 per cent of whom are women. Huws suggests, however, that there will be a significant increase in the number of lesser-skilled operating new technology from home (telecommuting) in the future as the cost of communications drops. She points out that

[2] A full account of Huws' research on home-workers can be found in Huws (1984).

in the United States, where cabling is much more extensive, there is one data entry company with half its 1600 workers operating from home. Three of the largest ten US companies (Standard Oil, Atlantic Richfield, and IBM) have already experimented with telecommuting. So far, new technology home-workers (unlike traditional home-workers) are usually paid on a time rather than a piece-work basis (Huws' study reported 80 per cent so paid), but high productivity gains are reported. F. International claims a 30 per cent improvement in productivity in telecommuting compared to office working, and the British computer firm ICL, which uses 200–250 telecommuters, claims that '25 hours work in the home is equivalent to 40 in an office' (*Labour Research*, July 1984, 171), a productivity increase of 60 per cent.

Trade unions are very concerned about the spread of home-working, both in general and in the new technology sphere, but they do have considerable difficulties in opposing it. Most of their members currently employed in offices are likely oppose the job losses implied by moving to telecommuting. However, ex-union members now at home with small children may welcome the opportunity of telecommuting. It is physically very much more difficult for unions to recruit, organise, and mobilise telecommuters as they are so dispersed. It has also been the case in the past, with traditional home workers, that there has been a very low level of union membership partly because home-workers are liable to instant victimisation and they cannot claim unfair dismissal without proving employee status, which is usually either difficult or impossible. The situation is exacerbated in Britain which has, apparently, 'probably the worst legal protection for homeworkers of any developed country' (*Labour Research*, July 1984, 172).

For professionally qualified telecommuters the question of union representation may not be uppermost in their minds. It is, though, interesting that in response to an enthusiastic write-up of F. International in the *Guardian*, portraying it as a socially progressive firm whose operations are showing the shape of things to come, one of their ex-consultants wrote to the newspaper detailing the rates of pay available through F. International, claiming that they were half the rate paid for full-time office-based workers doing similar jobs (*Guardian*, 23 Apr. 1985).

Swedish unions, in contrast, appear to have got a great deal further both in the thinking that they have been doing about the longer-term effects of new technology, and in the legislation that has been enacted.

Within the context of the Co-determination Law of 1977 the blue-collar trade union confederation (LO), the private-sector salaried employees' union (PTK), and the employers' federation (SAF) reached settlement in 1982 on a development agreement covering a number of items of specific relevance to the introduction of new technology. For example, it stipulates that development work for job organisation should aim to create varied and fulfilling jobs which will help employees to broaden their field of knowledge and experience and thus equip them for more demanding and responsible tasks. The agreement states that team work, group work, and job rotation are examples of ways in which development can take place in the job context. Moreover, the agreement means that employees' potential for acquiring increased proficiency and assuming greater responsibility at work are supposed to be taken into account when technical changes are made. The agreement gives employees the opportunity of improving their professional skills, and the employer is to provide training in new work processes which technological change may entail.

More far-reaching provisions of this agreement give the union the right to use consultants, paid for by management, to advise it on local new technology applications, and, if new technology is introduced in one plant in a concern, then the union has the right to involve in consultation at that time all those likely to be affected if the set-up is to be used in other plants in that concern.

The LO has backed up this agreement by setting up a system of 'study circles', involving an estimated 31,000 workers. These circles have been used to consider new technologies specifically, and the LO has also run one-week courses.

The white-collar workers' confederation (TCO) has also been active. It, too, has a study circle movement, and has produced two booklets on new technology for the study circles to consider. The TCO has sponsored research by a number of academics into such topics as the impact of CAD/CAM and into the question of home-working. One of the TCO's member unions, the ST (Union of Civil Servants), drew up in March 1982 a *Programme of Data Policy for the Union of Civil Servants* which contained, among many other items, the following demand:

All forms of individual performance measurement concerning, for example, working pace, the duration of breaks or the number of mistakes made, must be prevented.

Activities must not be fragmented in such a way that certain duties (data punching and data input) are turned into domestic employment via terminals or external typing pools.

The trade union organizations must have the last word as to which data are to be collected concerning employees and how those date are to be used.

One absolute condition for our acceptance of reorganisation and the introduction of new technology is that they satisfy the demand for less division of labour. (ST 1982a.)

A number of new technology agreements have been signed between ST and various state enterprises. Though these have not incorporated the demands as set out above, one of them does, for example, say that:

The aim of all computerization must be for systematic solutions to be designed in such a way as to create interesting and variegated duties for the users of the system. Terminal work must be organized and work points designed in such a way as to comply with current directions and recommendations, i.e. so as to achieve a good working environment. (ST 1982b.)

It is very tempting to conclude that one important factor leading to the Swedish trade unions' faster progress in responding to new technology is the form of trade union organisation in that country. It was suggested at the beginning of this chapter that some of the major impacts of new technology are to reduce plant size, increase the diversity of employment contracts and payment systems, and create major changes in skills required (thus changing the relative importance of various occupational groups). Unless there is a very high level of unionisation, backed by legal or other binding agreements guaranteeing the place of unions in the bargaining process, (and Sweden fulfills both these conditions) then unions based on occupational groups, with multi-union workplaces, are going to be at a disadvantage compared to industry-based unions.

This points to an acceleration of the trend in Britain now, which is that management offers just one union the bargaining rights for all workers in a particular plant. This has now happened on a number of greenfield sites where Japanese-managed companies have set up business, and the most striking example to date, as this book goes to press, has been the agreement between Rupert Murdoch of *News International* and the EETPU with regard to the manning of the new Wapping complex where the *Times, Sunday Times, Sun,* and *News of the World* newspapers are produced and printed.

There are, of course, a variety of alternative organisational responses which trade unions could make to the new technology, other than allow one union a monopoly of bargaining rights in a particular location. These would all involve closer co-operation between existing unions and, possibly, stronger central co-ordination by the TUC. There is as yet, however, not very much evidence that a response of this kind is being developed.

9
Implementing New Technology

We have in previous chapters discussed the many factors influencing the relationship between new technology and work organisation, and in the preceding chapter we examined the response which trade unions have made to the implementation of new technology in the work-place. In this chapter we review the various techniques which managers are recommended to use in implementing technical change and then examine some examples of different approaches to implementation, taking our evidence from the installation of new printing technology in Fleet Street and new car body assembly technology at British Leyland in Longbridge, Birmingham.

We then go on to suggest a new way of considering what techniques might be used in a particular situation, arguing that there are a number of contextual factors which need to be considered. What may be appropriate in one context may be quite inappropriate elsewhere. We specify what we think some of these factors might be, setting out a framework which managers and trade unionists can use to check the appropriateness of the implementation technique being applied in their work-place.

What managers are recommended to do

A particularly comprehensive and well-thought-out guide for managers on the implementation of new technology has been produced by the Institute of Personnel Management (IPM). Called *How to Introduce New Technology: A Practical Guide for Managers* (IPM 1983), it provides guidance based on a survey conducted by the Institute of practices and attitudes of firms who had successfully introduced new technology of one sort or another into their work-places. The first thing they emphasise is the need to plan the change. The IPM claim that if there is no planning then there is likely to be failure to:

Secure trade union commitment to the change.
Secure trade union agreement to operate the new equipment.

Make adequate plans for training employees to use the new
equipment.
Make adequate plans to avoid compulsory redundancy.
Recruit people with the skills necessary to operate or maintain the
new equipment.

A common problem with the investment in new projects is
that decisions are made initially by technical and finance staff,
so that by the time the personnel function becomes involved it
is rather late to make the kind of plans needed to avoid the
kind of failures described above. To get round this problem the
IPM put forward the interesting suggestion that there should
be a 'sign off' system for every investment proposal such that a
senior specialist from each affected function in the enterprise
has to sign the investment proposal before it can proceed. With
such a system, then, the signature of the personnel specialist
should be required, in addition to those of the technical and
finance functions, before any investment proposal begins to be
implemented.

With the personnel function involved at the outset, action
plans can be drawn up to include the following topics: training
and manpower planning; job design and work organisation;
employee relations; pay and conditions; and health and safety
issues. The next question posed by the IPM is who should draw
up these plans. The choice, they suggest, lies between setting
up an inter-functional management team or delegating the task
to a specialist management function.

If the former option is chosen, then the team should usually
include technical, personnel, and finance people together with
representatives of those line managers who will eventually be
directly involved in managing the new system. There is a well-
documented study of the planning and commissioning of a new
Courage's brewery using novel microprocessor-based production techniques (Work Research Unit 1982) where this
inter-functional planning team technique was used. Shell Petroleum also used this method in planning the introduction of
word processing into their enterprise (IPM 1983).

The alternative strategy of using one specialist management
function to do the planning usually involves either the work-
study or management services department, and the evidence
seems to be that success is less certain. The obvious advantage

of the specialist approach is that it will often be far easier to co-ordinate members of the same department rather than having to operate across departmental barriers. All will be technical specialists and therefore there will be no time wasted in explaining what is technically possible and not possible. The chance of misunderstandings is likely to be reduced. Equally, the obvious disadvantage is the rather narrower perspective that the single-discipline team is likely to take. The City of Bradford Metropolitan Council asked their Management Services Division to investigate the potential of word processing with the result that clerical workers came out on strike. A subsequent enquiry by the council concluded that in future the personnel function should be fully integrated into the planning process, there should be joint agreement between management and unions, and there should be full prior discussion with employees.

Thamesdown Borough Council, on the other hand, successfully introduced word processing on a basis planned by their management services function, but in this case the personnel function was involved and the local NALGO branch representatives were consulted at an early stage. When agreement was reached on a number of key issues such as salary grades, the reduction of posts through natural wastage, and the method of selecting the word-processor operators, there was a major communications exercise to inform all affected staff about the implementation of word processing. The IPM's conclusion seems to be that either inter-functional or single-functional planning is acceptable provided that the personnel function is involved enough to be able to ensure that the 'people' side of the technical change is adequately thought through. Even allowing for the likely bias of the IPM in this matter, it does make sense to suggest that someone involved in the implementation of new technology should have some awareness and skill in dealing with the kind of issues identified by the IPM and listed at the beginning of this chapter.

Once the team responsible for steering through the innovation has been set up, one of its tasks is to prepare the policies relating to the personnel aspects of the innovation (the ones listed by the IPM are: training and manpower policy; job design and work organisation; employee relations; pay, benefits, and conditions; and health and safety aspects). Of particular relevance to the theme of this book are the comments of the IPM about job design and employee relations.

What they have to say on job design is of especial interest. Bear in mind that we are reporting the views of the personnel function, and this may well not represent the ideology of line management in an enterprise. However, line management are not themselves represented by an official organisation which is setting out a coherent view on how to treat the 'people' aspects of technical change. So what the IPM says represents the major articulated position on job design from the management side, and it is unequivocally anti Braverman and anti de-skilling in its tone. The IPM argues that policy on job design and work organisation needs to take account of two main factors: the needs of the organisation for greater flexibility in manning, working methods, and operational efficiency; and the needs of employees for some degree of job satisfaction as a result of performing varied and interesting jobs. In other words, they reject the strategy of job fragmentation, de-skilling, and the tightening of direct management control in favour of the responsible autonomy strategy. They do this on efficiency grounds: the organisation of the future requires flexibility and therefore the exercise of discretion and responsibility by workers in the enterprise.[1] On the question of employee relations, the IPM sets out a list of six possible strategies, which they describe under the headings below. Although they do not point this out themselves, their list begins with the strategy which safeguards managerial prerogatives to the greatest extent and ascends (or descends, according to one's view) to the strategy which is most radical in terms of management–union power sharing. The various strategies are:

introduction by management decision
communication
direct employee involvement
consultation
negotiation
new joint structures

[1] For a more detailed exposition of this point, illustrated by case material, see Kelly (1985).

It is clear that the IPM favours the middle four of these six options. Warning is given that simply to introduce new technology by managerial diktat, without any prior communication, consultation, or negotiation, is likely to lead to repercussions from the workers. Instances are quoted of new technology standing idle for months, because workers refused to work the equipment when it was installed until there had been proper consultation. Putting in new technology on the basis of managerial prerogative alone will only be successful, suggests the IPM, if the company is facing bankruptcy or times are seen to be particularly hard by all the workforce.

In their survey the IPM found that communication with the workers via briefing groups or communication groups was widely established and played an important part. So did direct employee involvement in the form of feasibility study groups, discussion groups, and visits to suppliers or other users. They also discovered extensive formal consultation at one or a number of stages prior to implementation. None of these three forms of relating to affected employees cuts across what management generally regard as their prerogative, and in communication, consultation, and involvement the initiative remains in the hands of management. All three strategies are strongly endorsed by the IPM. The point at which they become much more cautious is when negotiation begins. They point out that there are two ways of handling negotiations over the introduction of new technology. One is the informal approach relying mainly on consultation procedures, and where management negotiates only across a narrow range of issues. The other is the attempt to reach a formal and fairly comprehensive written agreement, usually in the form of a specifically named New Technology Agreement (NTA).

From the arguments set out by the IPM it is easy to draw the conclusion that they would rather not have NTAs if they could get away with it. They point out that in the service sector of the economy, where trade union pressure is recent, management have resisted any negotiation over technical change as such, arguing that to do so concedes the principle of the status quo. This 'status quo' principle, otherwise known as the mutuality principle, states that there shall be no change in terms and conditions of employment, or in working practices, until such change has been agreed between management and the appropriate union representative. This principle has been generally accepted for decades in the, strongly unionised, engineering industry and, as we shall describe later in this chapter, BL management, to take one example, fought hard and successfully to remove it. The absence of the mutuality, or status quo, principle gives management much more freedom to introduce technical change without negotiation if they think they can get away with it. To accept an NTA which would state that all technical change can only proceed after full consultation, negotiation, and agreement would be a restriction on managerial freedom that most managers would only accept if forced to.

If management cannot keep themselves entirely free from negotiation then their next best preference would be to use existing procedures. This allows them not to accord to new technology any special status, again giving more freedom than the third option of accepting the separate NTA. Though they recognise that the existence of such a document may provide an orderly framework to the bargaining over new technology, the IPM admit that such agreements are usually only set up because of trade union pressure. They are found where white-collar trade unionism is long established, for example in the engineering industry. Of fifty-four NTAs surveyed by the IPM, 37 per cent were in engineering.

Where NTAs have been made, many provide for the establishment of what the IPM rather coyly terms 'new joint structures' of employee involvement. Other terms to describe these mechanisms would be 'participation' or even 'industrial democracy'. Over one-third of eighty-six NTAs surveyed by Williams and Moseley (1981) provided for trade union involvement in decision making prior to implementation. Seven involved the unions in the earliest planning stage (though all were in the service sector and four were in the public sector). In twenty-three agreements there was trade union involvement in selecting options for change and discussing the results of feasibility studies.

The IPM document cites three examples from the Williams and Moseley study, all involving agreements with ASTMS. One, with the Scottish Provident Institution, sets up a 'Productivity and New Technology Committee' which has equal management and trade union representation though it is chaired by management. It has the power to set up joint departmental working parties. An NTA with the Royal London Mutual Insurance Society sets up a Management steering committee which includes two trade union representatives and evaluates possible applications of technical change following initial research or feasibility studies. Proposals from this steering committee are passed to a 'Technology Conference' on which sit equal numbers of management and trade union representatives. Finally, Rolls Royce at Derby have an Eastern-European-sounding 'Central Committee' comprised of the company's systems management and trade union representatives, and chaired by the Staff Employee Relations Manager. Though it only meets four times a year, it is supported by a second tier of local discussions between management and the trade union representatives in the area concerned.

Despite the existence of these various new joint structures, and at least 225 NTAs by 1982, a Confederation of British Industry (CBI) survey of that year discovered that in relation to 'technical change' 44 per cent of units surveyed 'communicated', 42 per cent 'consulted', and only 9 per cent 'negotiated'. In larger plants negotiation was more common (presumably reflecting a higher density of union members), but even among units of over 1,000 employees only 16 per cent entered into negotiations over technical change.

It is interesting to note that the CBI itself very nearly agreed a joint statement with the TUC about the advisability of NTAs. The proposal for these agreements first became prominent following the publication in 1979 of the TUC's *Employment and Technology* document. The CBI responded with *Jobs—Facing the Future* (1980), and it was on the heels of this that a joint statement was drawn up. Though the TUC ratified it, the CBI did not. They agreed on the worth of taking a planned approach to technical change; the need for the disclosure and communication of relevant information; joint consultation and joint implementation; attention to health and safety matters; the need for training, retraining, and redeployment; and attention to job design to avoid routinisation of work. The CBI disagreed with the TUC, however, on the principle of formal agreement prior to the introduction of new technology (the 'mutuality' provision); the CBI also felt that employers should not be expected to provide job security guarantees as a quid pro quo for employee co-operation; and they took the view that linking shorter working time to the introduction of new technology would set an unacceptable precedent.

Some examples of attempts to implement technical change

The introduction of robots and multi-welders at the big Austin Rover assembly plant at Longbridge in Birmingham illustrates very graphically many of the issues we have discussed in this book about the process of introducing new technology.

The story begins in 1974 with the collapse of British Leyland and its rescue by the government. The committee set up by the government to advise it on what to do with the crashed company recommended, among many other things, that £1.5bn. should be spent in capital investment on car operations; that the product range should be greatly reduced and rationalised to cover just five model ranges; that the production facilities should also be greatly rationalised; and that among other specified reforms to labour relations some form of industrial democracy should be introduced.

Before the company collapse, there had already been discussion about a replacement model for the smallest, and most successful, car produced by the corporation—the Mini—which had been in production for over a decade. After the rescue, the development of a new model for this market niche became the company's highest priority and 'the yardstick by which the whole company would be judged', to quote the new Chairman of the company, Sir Michael Edwardes (Edwardes 1983, 178). The Metro, as the Mini replacement was to be called, was eventually launched in October 1980 and has enjoyed considerable market success. To manufacture the new car British Leyland invested heavily in new, microelectronics-based, production technology and at the same time attempted to introduce a number of innovations in work organisation. How these technical and social innovations were implemented we now briefly discuss.

Of particular interest to this story is the wax and wane in the use of industrial democracy in the company as the Metro project was developed. By October 1975, as many of the crucial decisions about the new production facilities were being made, a scheme of worker participation had been agreed, in line with the recommendations of the Ryder Report. This involved the setting up of joint management–union committees at corporate, divisional, plant, and unit level in the company and, of particular interest to our story, sub-committees of these joint committees at corporate and plant level to deal specifically with the introduction of the Metro (code-named the ADO88/LC8). The Participation Scheme is shown diagrammatically in Figure 9.1. The operation of the scheme is described in greater detail elsewhere, in a book-length study reporting research done by Paul Willman, Graham Winch, Mandy Snell, and myself (Willman *et al.* 1984). What follows is just a brief outline.

The Participation Scheme had, at least in the early stages, a considerable measure of support from management, the unions, and the government. The Labour Government of the time was strongly committed to tripartite arrangements with the trade unions and management and to the policy of introducing industrial democracy in industry. Management was not opposed to some form of participation and, indeed, participative arrangements had existed in BL in different ways in different plants for some years prior to the Ryder Report. In fact, in 1974 BL management themselves set up a working party to discuss the rationalisation of these decentralised arrangements (Davis 1979, quoted in Willman *et al.* 1984). Moreover, the trade unions had put forward a set of proposals to the Ryder Committee setting out a scheme for full democratisation at Board and plant level. It was generally accepted within BL that a 'quick implementation of the Ryder proposals was . . . essential to demonstrate that the Company had a future and that the unions had a role in it' (Willman *et al.* 1984, 87).

However, the evaluation to determine the most appropriate technology for the Metro took place in 1974 and 1975, before the participation scheme was set up. The major technological development in car production at that time was the increased use of automated welding of the car body. Until the early 1970s, most of the 2,500+ welds necessary to put together the metal

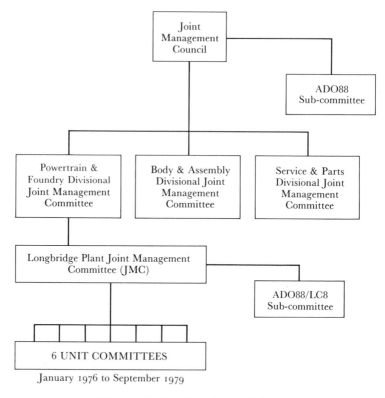

January 1976 to September 1979

Fig. 9.1. BL Cars Participation Scheme.

body of an automobile were done by hand-held welding guns. With developments in microprocessor control it had become possible by this time to perform welding operations with a combination of multi-welders and robots. The main difference between multi-welders and robot-held welding guns is that the former are a less flexible form of automation comprising a bank of welding guns each of which puts its spot weld in just one place. A robot-controlled gun can be programmed to perform a sequence of spot welds at a variety of points on the car body. It was the introduction of automation to produce the car body that represented the major introduction of new technology to the Metro production process.

The decision to follow the high technology route was taken on three main grounds. It was seen to be the technology of the future and it was believed that the company ought to get into it sooner rather than later. Secondly, it was felt that automated welding would give more consistent high quality than manual welding. A major problem with manual welding is ensuring that the body is fitted together precisely, that all welds are actually made, and that the welds are positioned accurately. Automated welding was believed to be both more reliable and to give feedback if a welding operation failed. As the quality of the finished BL products had been widely criticised in the past, this second factor was of considerable importance to the company. Thirdly, though the company could not make the overall cost of using the high technology look cheaper than more conventional assembly, there were felt to be un-quantifiable advantages in using a technology which resulted in management having to deal with far fewer direct operatives. It was estimated that the saving in direct labour using this high technology production process would mean a cut by two-thirds in the amount of direct labour required compared to that which would have been needed with conventional 'gateline' tech-nology. Put another way, an extra 942 jobs would have been required if the conventional body build arrangements had been used. (Willman *et al.* 1984, 57 and 60). A possible fourth reason was the willingness of the government to fund capital invest-ment. BL may have felt its future more secure with capital investment than with a solution which could have resulted in subsidies paid as wages during the production run.

The decision about which technology to adopt, together with estimates about associated productivity, were enshrined in a company document known as the 'Red Book'. This was prod-uced in 1976 by the manufacturing engineering function just as participation was about to begin and made up a major item on the agenda of the ADO88 Sub-committee of the Joint Management Council. This sub-committee met frequently, working to a very tight timetable, between April 1976 and May 1977, and, because of the strict timetable, management decided to merge the normal management decision-making process with the sub-committee's business. Thus the union repre-sentatives on the sub-committee were invited into weekly manufacturing planning meetings.

The sub-committee made three reports to the Joint Management Council and these reports accepted the Red Book proposals for the high technology route and for productivity targets. In other words, proposals to introduce new technology with striking implications for job losses were accepted almost without question by trade union representatives in a company that had up to this time been a household word for trade union militancy. It is hard to assess to what extent the vulnerable financial state of the company persuaded them to agree, and to what extent their involvement in participation was an important factor. The conclusion in our book is that:

Although both the management and trade union members of the Council sub-committee interviewed stressed the free flow of ideas that took place throughout, they felt that the unions had little impact on the overall shape of the ADO88 project. The participation of the unions in the planning process merely led the unions to accept the recommendations of the Red Book almost wholesale, and in particular the important clause on productivity targets. (Willman *et al.* 1984, 96.)

In addition to choices about technology and productivity targets there were two other issues which are worth discussing— job design and work organisation. In 1973, as part of their interest in the issue of participation, BL recruited several personnel staff to review ideas on work organisation. Willman *et al.* describe how, during 1974 and 1975, when decision making on design and technology was taking place at Cowley, factors such as length of job cycle times and their effects on workers' output, job satisfaction, and job rotation were brought into the evaluation of alternative body framing systems. A systematic analysis of such factors resulted in a 'rating' of alternative systems on a combination of engineering and 'quality of working life' criteria. Though a Volvo-style method of working was rejected as unsuitable for the high-volume Metro operation, an alternative system of 'team-work' was considered. This was a system of flexible working which made a group of workers responsible for production and quality within a given area. The team members themselves were to decide how and when job rotation would take place, subject to the productivity constraint.

In the event, for reasons which will be described shortly, these considerations ultimately exerted little influence on the final design of jobs. Eventually the option was for very short cycle times without team-work on the Metro line. Though a team of BL managers and trade union representatives had visited the SAAB plant at Trollhatten where identical automated welding machinery was already in use, none of their work organisation principles were adopted. Instead of the SAAB line-out system, described in Chapter 5, and use of carousels on the multi-welders in order to eliminate machine-paced short-cycle jobs, BL adopted 'single lines with very short cycle times and low job satisfaction' (Willman *et al.* 1984, 55).

There were two major proposed changes in work organisation. The one was to reduce demarcation between the maintenance trades down to two major categories of work—the proposal that became known as 'two-trades maintenance' and which is described in Chapter 5. The other was to introduce a form of team-working on the shop-floor.

The proposal for team-working was to organise operations within a given production zone such that there would be little demarcation between jobs of particular members. Functional foremen would be replaced by team foremen responsible for supervision of the multi-functional zone team, which would comprise the operators, a quality controller who would both inspect and rectify, material handlers, and on-line maintenance personnel. Within the teams, it was proposed that job rotation would take place, and also that 'progression' from material handler to operative to quality controller to maintenance would be possible. Moreover, the operator was to be responsible for 'minor' maintenance jobs such as air and water leaks, and initial fault diagnosis.

Though these proposals were accepted by the joint trade union– management ADO88 Longbridge sub-committee, a significant number of managers opposed team-working. Several objections were advanced: the company did not possess the managerial resources to supervise the new scheme; job rotation was likely to lead to a problem in assuring quality; and any advantages accruing to labour flexibility were overstated.

Nevertheless, the proposals for two-trades maintenance and team-working became official management policy for a time, and management began to negotiate their introduction with the unions. By this stage of the proceedings, however, BL was experiencing its greatest financial difficulties yet, the direct cause of which was the continuing plummet of its market share to below 20 per cent of the UK market. Following the appointment of Sir Michael Edwardes as Chairman, and the general tightening of management control in the company, the unions withdrew from the participation scheme and management changed their tactics with regard to negotiation. The eventual result was that a document detailing changes in working practices was unilaterally imposed on the workforce. The company gave notice that the proposals of this document, known as the 'Blue Newspaper', 'would be implemented on 8 April 1980: any worker arriving for work on that day would be deemed to have accepted the package' (Willman *et al.* 1984, 134).

The result by the time the Metro was launched in October 1980 was that, although two-trades maintenance was never agreed, a compromise on two-trades response to breakdown was achieved for the new line. With regard to team-working, the full proposals were never implemented, but there does seem to have been a development towards some form of team-working with BL recently. However, on the Metro line, the pattern of work organisation is dominated by the extent to which managers control, on a daily and hourly basis, which machines are to be operated. Because the productivity of the new technology was even higher than forecast, and because sales of the Metro have never been as high as total planned capacity, the various sections of the automated lines do not have to be used continuously on both shifts. Thus management allocates labour to particular parts of the line at various times of the day in order to run the plant with fewer operators than would be required to keep all the machinery operating all the time. Thus stable patterns of team-working cannot be established.

We have seen how new technology was introduced onto the Metro line using quite extensive participation methods. Let us contrast it by describing briefly the Fleet Street experience in introducing new technology—or, rather, failing to do so.

Anyone with access to a personal or home computer and a cheap dot matrix printer has some awareness of the dramatic changes which new technology has made to the production of the printed word. In newspaper production the potential impact has been tremendous. Until recently, the newspaper production process involved four transformations of the medium on which the message to the newspaper reader was conveyed. Journalists would type out their stories on paper, to be processed by sub-editors. The typescript would then be set up in lead by the printer. From this leaden image the printing plate would be made, and this, when placed in the printing press would, produce the final product. The new technology has halved the number of transformations needed. If journalists write their stories onto terminals, so that their input is stored in electronic form, then sub-editors can work from this input and the printing plates can then be made from it by a process known as photo-typesetting. Performing the production process this way offers advantages both in direct labour saving and in reducing the lead time from input of copy to production of the printing plate. An additional benefit to management is the opportunity it gives them to weaken the central position of the craft printers in the process. Printers have historically had extremely strong control over their own occupation, through their union, the NGA, and in Fleet Street their position has been pre-eminent.

In the provinces many local newspapers have successfully transferred production to this new process, often with very little conflict. In Fleet Street, however, the union successfully opposed the full introduction of this new technology until early 1986, when Rupert Murdoch's News International company (which includes the Times, see below) set up their new printing works in Wapping. Their struggle to do so has been described in some detail by Roderick Martin (1981) and here is a summary of the tale.

The three newspaper companies Martin discusses are the Financial Times, Times Newspapers, and the Mirror Group Newspapers. At first the employers attempted to set up a Joint Standing Committee comprising the Fleet Street newspaper proprietors and the unions, but failed to do so. The Financial Times then went it alone with direct negotiations. One of their major objectives was to get the NGA to agree to allow the journalists and other non-print union staff to input their own material directly into the photo-typesetting computer. An equally major objective of the NGA, and one which they have successfully maintained to this day where unionised printers are employed, was to maintain union control over input to the computer. When the Financial Times failed to reach agreement with the NGA over direct inputting (journalists etc. putting their own material directly into the computer) they withdrew their proposals to adopt the new technology and instead adopted the quite different corporate strategy of launching a European edition of the paper, printed in Frankfurt.

Times Newspapers Ltd. adopted a much more aggressive confrontational style of attempted implementation. They actually bought the new equipment and installed it ready for use, while concurrently negotiating with the unions. Management was also attempting to negotiate with the unions across a range of other issues at the same time, and, as a result of failure to reach agreement on these, publication of the various Times newspapers was suspended in November 1978 for a period of eleven and a half months. After resumption of publication no newspaper or supplement was produced on the new equipment while Times Newspapers Ltd. was under Thompson ownership.

The Mirror Group had more success, though this was limited. They got new technology into the building, and got people to operate it after a fashion, but many of the facilities of the new equipment could not be used for technical reasons. Management's intention was to use the equipment to set up the entire layout of each page of the newspaper so that once the written material had been entered into the computer it could be manipulated and structured into final page layouts. For a variety of technical reasons, this turned out to be impossible, and so the equipment was being used, at the time of Martin's study, only to produce individual articles from the phototypesetter which were then made up into pages by cut-and-paste techniques. These would then be photographed to make the printing plate.

Martin suggests a number of reasons why these major Fleet Street newspaper groups were unable to implement the new technology successfully. One of these, which contrasts nicely with the BL case discussed earlier, was the financial position of the newspapers. At the time of these first attempts to bring in new technology, the financial position of Fleet Street was thought to be extremely precarious, with a Royal Commission examining the future of the industry. This had improved by 1980 and anyway, as events at Times Newspapers indicated, the unions appeared to believe that Fleet Street could bear the costs of not adopting the new technology. Other reasons, suggests Martin, were that, compared to the printing industry in the United States, the UK newspaper industry was inadequate with regard to the equipment it bought, the back-up facilities it could obtain, and the skills to manage it. We complete this section by discussing two further implementation strategies. One is the attempt to stimulate innovation from the centre of a decentralised and diversified corporation. The other is the setting up of new technologies on green-field sites.

It is a characteristic of many British companies that they are, by international standards, very large, but comprised of individual plants which are, by the same standards, rather small. This is especially true in the engineering industry. This can cause a major difficulty in taking up new technology because there is often little expertise available at plant level about technological developments, while at corporate level there may be too little knowledge about the particular requirements at the local level. This was illustrated in one large corporation which had a large number of subsidiary companies across a diverse range of industrial sectors, some of which were in engineering. The decision was taken at corporate level to encourage companies within the enterprise to take up new technology, and especially CAD/CAM where it was appropriate. In order to encourage such innovation the headquarters set up a separate development company, within premises already owned by the corporation, with three main functions. It was to act as a demonstration centre for this technology for the group of companies. It was to operate on a bureau basis, doing CAD/CAM work under contract for other companies in the enterprise and elsewhere, and it was to provide an advisory service for other subsidiaries which were contemplating the setting up of their own facilities. At the same time the technical director at headquarters was encouraging the chief executives of the various subsidiary companies to invest money in CAD/CAM. Though the new development company clearly did some valuable work, it seemed that the pressure on individual company-level chief executives led some of them to install equipment which turned out to be unsuitable for their particular applications. One particular company had two toolrooms and bought, as a result of this corporate-level exercise, two CNC machine tools. Neither was appropriate to the toolroom it was put in. One was blacked by the local union and stood unused for several months; the other had the bed of the lathe broken through misuse within a week of installation. On the other hand, at much the same time, this company also investigated, on its own initiative, the possibility of installing a computer-based management information system. It hired local consultants to mount a study, and to design and install a system. This appeared to be accepted by the end-users with little difficulty.

Another method of implementing new technology is to transfer operations to a green-field site. The advantages to management of this strategy are that old working practices are not transferred across to the new equipment; different methods of payment and management control can be adopted; and a new workforce can be recruited where management may be able to exercise some control over who joins the firm. The disadvantages of this strategy to the unions are of course the obverse of the advantages to management. The BL Metro production facility described earlier is an example of this being done on a major scale. The new facility was set up on a new site opposite the old works in Longbridge and BL management took great care to allow to transfer from the old site only those who had what they termed a 'positive attitude to work'. Management also took some care to arrange the type of union organisation that developed in the new plant.

Another multi-site company which wished to develop the use of CAD/CAM took the strategic decision to introduce the technology in an intensive way on just one site in the first instance. Instead of setting up from scratch a green-field site they chose a small plant located in a rural area, using non-union labour. They were able to introduce both CAD and CNC machine tools without any formal negotiation and operated an extremely flexible system of work organisation. They were thus able to experiment with different ways of using the technology, and were also able to set productivity standards which could be used in negotiating the introduction of the technology in their unionised sites in the Midlands.

Designing new technology systems: the options

In the first part of this chapter we reviewed the IPM's advice on how to introduce new technology, and we have just reviewed a number of detailed cases of how firms actually went about it. Our conclusion is that there is no 'one best way' to introduce new technology, and in the section that follows we set out a methodology, first developed in a book on office automation and organisation jointly authored by Judith Wainwright and myself (Wainwright and Francis 1984), for assessing the costs and benefits of introducing new technology in various ways in different kinds of enterprise. Two major questions to be resolved in deciding how to implement new technology are 'Who should participate in the design?' and 'Where should the decisions be made: how centralised or decentralised should the decision making be?' We can set up a matrix, as in Figure 9.2. The sides of the matrix can be thought of as continuous axes, with a broad spectrum of possible approaches between the totally centralised and totally decentralised, and again between having users as the main participants at all stages and having no user participation in any of the stages. These two dimensions are an alternative way of expressing the range of possibilities suggested by the IPM. Box W in Figure 9.2 represents the option they describe of using the single-functional specialist team for the planning, and simply introducing the new technology on the basis of management decision alone. Box Z at the other extreme would involve inter-functional teams and extensive user involvement in the design process, quite possibly incorporating the use of what the IPM term 'new joint structures' of decision making.

The style of implementation characterised by box W is the traditional computer systems design approach, employed since the earliest batch computer systems were installed in the 1960s. These were designed by data processing specialists in central departments who often had little contact with the rest of the user organisation. These systems typically automated highly standardised, routine clerical tasks, where the data obeyed a

Who participates

	All design done by computer specialists	Users participate in the design of systems
Centralised	W	X
Decentralised	Y	Z

Where located

Fig. 9.2. Possible Design Approaches.

series of definable rules. Examples are payroll and other accounting applications.

The reasoning behind this approach is clear enough: designing and programming computer systems requires technical expertise. The tasks to be programmed are simple, clerical routines. Therefore, the argument goes, technical specialists from the central data processing department should define the tasks and program the computer as this is the most efficient way to get results. Once the system is ready, user procedures can be adapted to fit in with the computer's information needs.

The effects of these systems tended to be to increase the degree of centralisation of decision making and standardisation of tasks, and many users surveyed in the late 1960s complained of the inflexibility of their computer systems. Furthermore, the systems did not all increase efficiency or lower costs. In many cases the computer system imposed so many operating constraints that people spent even more time collating information to meet processing deadlines than they had previously spent doing the work the computer was meant to be making more efficient.

In addition, these computer systems seemed to decrease the quality of jobs of end-users by reducing the desirable 'socio-technical' design factors of the work such as those described in Chapter 3. Harry Braverman is not alone in reporting the tendency of computer systems to decrease the variety, responsibility, autonomy, and creativity of much clerical office work, and to make it more like factory work. A recent collection of studies of women working with computers comes to the same conclusion (West 1982). However, the apparent de-skilling of much office work was not the end of the story. In many cases, adverse reactions to such job changes on the part of system users have led to severe problems. Users may have great difficulty in getting so-called 'simplified' systems to perform the required tasks, or they may object to the changes in their work pattern so strongly that they find it impossible to work properly. They may even reject the new systems outright. Failure to meet human job needs has been cited as one of the most frequent causes of computer systems failure in the past (Bjorn-Andersen, 1976).

We can thus see two types of deficiency in many computer systems which have been designed using this traditional approach:

1. A system may not be appropriate for the task in hand—for example by imposing rigid standards and procedures which do not fit the task, and thus creating extra work or reducing flexibility to provide the required service.
2. The system may make work more unpleasant and/or less satisfying for the users than previously, by removing areas of discretion, variety, and interest. This, in turn, may decrease users' motivation to produce work of the required quality, or, more importantly, to make the system a success.

It seems highly probable that many of the adverse human outcomes and inefficiencies described above result from limitations in design objectives, resulting from the predominance of designers' and managers' values, which are oriented towards a narrow range of problems. It follows that some means to counteract this predominance is required. One way to do this is to increase the influence of user departments over the design process. Trends in computer systems design of recent years appear to show an implicit recognition that there have been considerable deficiencies in the traditional approach. This recognition may have been spurred on by today's spreading practice of providing data processing and information systems as a chargeable service to user departments. This can mean that, in theory at least, if the users do not like the system they can stop buying it, thus placing data processing people under more pressure to deliver what the users want at a reasonable price. The trends which are most significant for our purposes are:

1. It is now common practice for systems analysts to 'consult' users in order to 'define their needs'.
2. Some decentralisation of design activities sometimes takes place, with departmental or location specialists appointed to serve the needs of particular groups of users. An example is the employment of business analysts by financial departments.

These trends can be seen as movements downwards and to the right in the design approach matrix, resulting from the failure of a traditional approach to produce computer systems which are both acceptable to their users and appropriate to the increasingly complex and diverse tasks to which they are applied. However, it is not yet common practice for these movements in design approach to represent a major change in design practice. The emphasis is on 'consultation' rather than 'participation' and on 'definition of needs' by technical analysts for users (since the former are supposed to know what is realistic technically and what is not) rather than on users' specifications of their own requirements.

Alternative design approaches

Just as organisations have a choice of technical systems to adopt, so it is possible to choose a design approach according to an organisation's needs. Indeed, the approach chosen is likely to have a major influence on the design decisions which are made, and hence on the ultimate choice of technology, application, work organisation, and so on. We turn now to the alternative design approaches which can be chosen, and the implications of choosing them.

Approach X: centralised participative design

This approach makes use of a centralised committee or group of user representatives to work with the systems specialists in developing the design. The joint consultative committees set up within BL and described earlier are an example of this approach. It is appropriate to systems where a degree of centralisation arises through users from different groups, or individuals from different areas, sharing common needs through similarities between the tasks which they perform. It allows for the automation of complex tasks, such as the body-welding line at BL Longbridge, or those where it is necessary to secure the commitment of users to making the system work once it is installed. Centralised design processes may be necessary in circumstances where there are various departments sharing a system, and where the work of each of these various departments depends quite closely on the output of other departments— what Thompson (1967) terms 'sequential interdependence'.

Approach Y: decentralised non-participative design

In this case, design activities are decentralised but specialists are employed to carry them out. These specialists might be seconded from a central data processing department or recruited by local management. An example of this approach is the employment of contract staff and consultants to set up a department's minicomputer. Here, decentralisation enables the designers to be more aware of local user and job needs, but the users themselves do not make a direct input to design decisions.

Where the needs of the task are localised, requiring adjustment to local environmental conditions perhaps, design approach Y becomes more appropriate. This depends upon local 'specialists', so it is important that inputs and outputs to the local system can be clearly defined. This is likely to be the case where interdependency with other organisational groups is sequential. Such an approach may be appropriate for some of the systems intended for use by managers and higher-level professionals. Their systems often need to be tailored to fit local and individual requirements, but managers may be unwilling themselves to participate in the tailoring process. This process could be perceived as being of lower status than their more usual activities.

Approach Z: decentralised participative design

Two examples of this approach can be seen in organisations today:

1. Users may have their own microcomputers or personal computing facilities on a central mainframe, which they use to build their own planning, modelling, or forecasting 'programs'. Such systems are normally used by highly qualified professional workers (for example, in investment departments of financial institutions) who are thus enabled to participate in the design of their own systems. Quite extensive design activities may have taken place without their involvement beforehand, in order to provide these flexible facilities.

2. The second example is the design of groups or departments of their own office systems. This has been done with considerable success by typists and secretaries for word processing systems. Either a formal mechanism can be adopted, as in Enid Mumford's approach to the design of an office system by secretaries in ICI (Mumford 1983). Alternatively, a less formal approach can be used by giving groups of users a 'skeleton' set of facilities with which to build their own applications. The latter was observed in one of the cases reported by Wainwright and Francis (1984).

Design approach Z is likely to be appropriate where commitment of the workforce to the implementation of a system to fulfil complex localised requirements is vital to the system's success. This is likely to be an important consideration when the new technology being introduced is going to be used by those people whose jobs are displaced by the old technology. Under these circumstances participation in order to gain commitment is important, whereas if the system is being designed for installation on a green-field site then this consideration is irrelevant.

Where those affected are in departments whose work is very interrelated, local participation may be the best way to ensure that the systems facilitate this. However, some central coordination is likely to be needed to ensure that different groups develop compatible systems and methods.

Where user knowledge and expertise is to be built into a system, a participative approach can ensure that the system reflects that knowledge accurately, that user interfaces are appropriate to their human and job needs, and that these same users are likely to be motivated to make the most of the system once it is available.

W	X	
Central control Economies of scale on specialists' time Simplified design task	More user commitment System more suited to job needs Economies of scale on specialists' time Socio-technical design	Advantages
Design not suited to complex work · Design not suited to user needs System may be rejected	Representation can be difficult More designer and user time needed Users need knowledge	Disadvantages
Departmental control of 'bought-in' expertise More responsive to local needs	User commitment Local task needs can be accommodated Socio-technical design Less specialist effort than Y	Advantages
Costly in specialist time as spread through departments May be insufficient user input to system	Depends on no threat to potential users' jobs Requires careful co-ordination Users meed knowledge	Disadvantages
Y	Z	

Fig. 9.3. Advantages and Disadvantages of Different Design Approaches.

Which design approach?

Each of the approaches W–Z has attendant advantages and disadvantages, depending upon the circumstances in which it is employed. These are summarised in Figure 9.3.

The purpose in introducing this typology is to illustrate that there are real alternatives to the type of design approach which has traditionally been used ·for computer systems. Moreover, the alternative advocated by handbooks for managers such as that produced by the IPM may well be better than the traditional approach, but it should not be taken as the 'one best way'. The choice over which design strategy to take should be guided by, among other factors, the kind of work being done in the particular situation and the characteristics of those workers likely to be affected by the technical change.

10

Conclusions

Much of what has been written about new technology at work has been speculative and some has been ill informed. Most of the early forecasts about the revolutionary impact of the microchip have turned out to be grossly exaggerated and there are, as yet, relatively few empirical studies of new technology at work upon which to build well-founded generalisations.

What I have attempted to do within this book is to gather together a representative collection of available research evidence about what is happening within enterprises as new technology is introduced, and to discuss these findings in the light of our theoretical knowledge about the behaviour of management, workers, unions, and organisations. This has been done partly to develop some forecasts about the likely effects of new technology at work, but the more important objective has been to clear the ground and to begin to develop a way of thinking about new technology at work that will enable practitioners, whether management or worker representatives, to become more aware of the variety of possibilities which the new technology makes available to them, and to be able to make intelligent choices from among those options.

Having said that, the first topic dealt with in the book, in Chapter 2, was the contentious one of the employment effects of new technology. The policy implications of this are much more a matter for government than for individual managements or unions at plant level, though representative management groups such as the CBI and unions at national level have taken a strong interest in the issue and published their own policy statements.

The most frequently encountered argument about new technology and unemployment is that with robots and other automated machines doing everything for us there will be nothing left for humans to do all day. This, I argued, was erroneous. Such forecasts have been made about the effect of large-scale technological changes in the past, most notably about automation in the 1950s, and have proved to be wrong, though it is accepted that because of its extremely low price and great versatility the microchip is capable of a much wider range of applications than any previous new technology and is therefore likely to have a much more pervasive effect on social and economic life than, for example, advances in automation three decades ago.

However, our conclusion that the argument was wrong was not based on the fact that the same forecast had been wrong in the past, but on the belief that most people had a long way to go yet before all their wants were satisfied. Compensation theory argues that, while people have money in their pockets and goods or services that they want, this demand will work its way through the system and generate new jobs to satisfy the unmet demand. Only when we have reached the point where we prefer, at the margin of our income, extra leisure to extra income, will the introduction of new technology in one sector not cause new jobs to be created in another sector. When that point comes, then those who want it will choose leisure, and a shorter working week. Those who still, at the margin, prefer the extra income should still be able to find work, provided that labour markets are working reasonably freely.

We also concluded, however, that compensation theory has its flaws. One problem is that, even if compensation theory works in the long term and the overall level of employment remains high, there may well be substantial transitional difficulties for individuals who are caught up in the process of technological change. Those worst affected would be people such as those with redundant skills and/or living in locations where towns had grown up around some particular geographical or geological feature that no longer has an economic significance. Miners and steel-workers are obvious examples. Policies which override market processes, such as regional development policies or ones which encourage firms which have declared large-scale redundancies to attract other jobs to the locality, are frequently being advocated and introduced in many OECD countries in response to these transitional difficulties.

A second problem with compensation theory, we concluded, was that a new, higher, equilibrium level of unemployment was likely to result from job losses due to technical change if an adequate level of unemployment benefit was paid without a corresponding level of reflation of the economy.

Thirdly, following and extending Leontief's argument, we suggested that the maintenance of full employment may require some degree of job subsidisation by government, a weak version of the Leontief argument suggesting moderate levels of subsidy. A strong and extended version, which emphasises the high fixed costs of employing labour which do not arise when capital is substituted for labour, would suggest that the societal optimum might only be achieved if government in some instances subsidised jobs to a level beyond the costs of the individual's direct wage costs.

Fourthly, the macro-economic argument about the instability of individual firms' investment behaviour was advanced. Firms may assume a growth rate of the economy below that which is attainable, invest accordingly, and cause the growth rate to fall by their own behaviour. This self-fulfilling prophecy is, presumably, even more likely to occur when potential growth rates are high because of the possibilities brought about by new technologies.

Fifthly, there is some evidence that technological development, and economic growth, does not flow smoothly and continuously but in jumps, or in long waves. If this is the case, though the economy and employment levels may cycle around some long-term acceptable average level, we should not expect a smooth transition in the adoption of radically new technologies.

The policy implications of our conclusions about new technology and employment are therefore as follows. At the level of the firm, managements may find it easier to achieve acceptance of new technology if they are able to make efforts locally to find new jobs to replace those lost. They might do this either by attempting to diversify the products or markets of the enterprise, by seeking to attract new firms into the area (the BSC Industries Ltd. scheme is a good example of this), or by trying to persuade local, regional, or national government to attract other industry into the locality. Unions at local level might adopt as one of their policies that of attempting to persuade local managements to adopt the above policies, and of approaching local government to encourage them to adopt job

generation policies. London, with its Greater London Enterprise Board, and Sheffield with a similar body are examples
of the kind of initiatives local unions might be seeking. The
policy implication for local government is that it is inadequate
to rely on market forces to attract jobs to an area at a fast
enough rate to maintain an adequate level of employment.
Local initiatives either to supplement or to replace the market
are necessary, though it goes beyond the scope of this book to
explore what has been, and could be, done in this area.

The policy implications of our analysis at national level are
implicit in what has already been said. Government support is
necessary to soften the blows of the transitional effects of technical change. Some degree of reflation and job subsidisation
either nationally or on a regional basis is necessary in order to
maintain the system at an appropriate level of employment.
The system fails if left to the sum of decisions made by individual
firms in the light of signals which the economy gives to each of
them as isolated units. The TUC, national unions, and the
CBI, on behalf of employers, all have an interest in pressing
government to adopt job support policies of this kind. Without
them, overall output from the economy will be less than it
would otherwise be and, because new technology will be causing unnecessary unemployment, it will be more difficult to get
it accepted within individual firms, thus putting an extra twist
in the vicious circle of relative economic decline.

In the rest of the book we turned to the effects of new technology on people's jobs and the organisations in which they
work. We examined the two extreme scenarios—the one predicting the proletarianisation of the workforce as new technology de-skilled work and the other predicting
professionalisation as new technology took over all routine
functions. In order to begin to understand the relationship
between technical change and people's experience of work we
then reviewed, in Chapter 4,the various theories and findings
about the influence of technology on job satisfaction. Our conclusion was that, although technology poses certain constraints,
how people react to it depends crucially on social rather than
technical factors. We showed, for example, that French oil-
refinery workers felt very differently from English refinery workers about their jobs, even though the process plants they were

controlling were rather similar. We showed that in some experiments designed to increase job satisfaction the level of satisfaction dropped because the workers in those instances did not want what management was offering them. Our description of the Tavistock studies, and the resulting socio-technical systems theory with its prescription of the factors that make a job a good one, showed that social relations at work were a vitally important factor influencing people's experience of work. Nevertheless, it is how such relations develop, and what they mean, that is at least as important as technical aspects of job design such as the degree of autonomy or responsibility in a job. Much of the rest of the book was concerned with the processes that go on as new technology is introduced, with a particular emphasis on the fact of conflict in this process and on how that conflict is handled.

Most studies in the 1950s and 1960s, including those done by the Tavistock Institute, assumed that there was an optimum form of work organisation and job design that would both give a high level of satisfaction to individual workers and a high level of productivity. Braverman's de-skilling hypothesis, and the historical and contemporary case material that came to be so well known during the 1970s, destroyed such complacency, and the issues of conflict and control at work became of central concern to those attempting to explain contemporary patterns of work organisation and job design. Of particular relevance to the theme of this book was the assertion that technology played a role in this conflict. It was suggested that it was a tool of management to enhance productivity, not just by increasing technical efficiency but by increasing management control over workers. This tighter control was used to force workers to produce more than they preferred to, and to do it for less pay per unit of effort. While recognising some truth in this analysis of the use of technology at work, we concluded that the truth was somewhat more complex and Chapter 5 gave details of a series of cases where new technology had been introduced into the work-place with consequential changes in work organisation and in people's experience of work. It was clear from these cases that the relationship between new technology and these other factors was actually variable and complex, with neither extreme scenario receiving support across the board. To explore and analyse this complexity it was suggested, in Chapter 6, that it was possible to identify four senses in which control is exercised at work. From this a typology of control was set up and there followed a discussion of types of organisation. We took a broad view and suggested that it was useful to consider three basic types of organisational arrangement. One was that based on the hierarchy of managerial authority, that traditionally found in most commercial and manufacturing organisations. Another was the peer group, a form of organisation jealously guarded by many professional occupations such as doctors, lawyers, and some branches of engineering. In those cases the form of organisation is termed 'a partnership'. Less august groups of workers might form worker co-operatives based on the same organisational principle. The third form of organisation is that of the market, used as an alternative to both hierarchy and peer group both by commercial enterprises when they sub-contract work and by partnerships if they buy in expertise in the form of consultancy. Having set up this typology of organisational types, we went on to examine how our four types of control were exercised in each. This, we felt, set out the range of options available and enabled us to ask what effect new technology might have in shifting organisation from one form to another. Would it, for example, lead to a withering-away of hierarchy and increased use of contracting? Would peer-group working become more possible, or comparatively more costly?

Having traced out these possibilities, and what preferences various interest groups might have for each form of organisation, we were then in a position to look at the challenge this posed for trade unions. Though many trade unions, particularly those representing white-collar workers, have developed policies about new technology, none of them have public policies about the kind of challenges to present forms of work organisation which we identified. The nearest any of them come is in the policies they have about job design. Some are renewing their concern about home-working, or 'telecommuting' as it is now called if the job is tied to the contractor by a cable link, but none have yet met head-on the kind of challenges that are about to be made.

These challenges, we conclude, are along the following lines.

In the past a great deal of economic activity has been co-ordinated by means of managerial hierarchies, such a form of co-ordination having been on the increase until very recently as large firms have continued to grow by merger. It is likely that such a form of organisation has been both technically efficient, for the reasons given by economists such as Oliver Williamson (1975) and by organisation theorists in general, and also effective as a means for capital to extract value from labour. For a variety of reasons, this trend may now be in reverse and we are seeing a switch from hierarchy to market. We see evidence for this in the number of dis-mergers, often by management buy-out, whereby firms are uncoupling pre-viously management co-ordinated parts of their enterprise, and in the very fast growth in the phenomenon of franchising, whereby a company establishes a market niche but supplies that market via a system of sub-contractors who own the retail outlets and serve the market under licence. Macdonalds and Hertz are two of the most common household names operating in this way. Some of the reasons for this may be to do with social factors which are giving a degree of power to groups in society other than those with capital. Particular professional skills, including those of administration, or entrepreneurial skills may now be a greater source of power than in the past. With widespread saving in society, whether through pension funds or in banks and building societies, there is a source of capital more easily available to those with these skills, so that the professional can rent capital, rather than the capitalist rent the professional. Thus the Marglin-type reasons (Marglin 1974) for hierarchy rather than market may be diminishing. Labour legislation and trade union power may also be a factor, re-ducing the capacity of capital to exploit labour within the employment relationship. Indeed it may now be easier in many situations for this exploitation to take place through the market place via sub-contracting than within the employment re-lationship. A further factor may be that, with enhanced tech-nological sophistication of products and processes, a richer and easier source of profits may be extracted out of the relationship between supplier and consumer rather than the relationship between capital and labour. Thus a fast food chain may be able to exploit the market created by its expertise in advertising, packaging, and distribution, and make profits from this more effectively than making surplus value from those who fry the chips and wash the dishes. In other words, there is more to be made from the percentage paid by the franchise licensee than from ensuring a high rate of labour productivity from the washer-up. The first and last of these factors are directly related to technological development. If technical development leads to higher qualifications and higher levels of real earnings in the working population at large then sources of capital become more widespread. With technological advance comes the possibility of exploiting particular market niches, and that, rather than the labour process, becomes *the* management process to concentrate upon. In both cases there is a reduction of the pressure on capital to retain managerial hierarchies as a means of controlling and exploiting labour. Sub-contracting, free-lancing, and use by companies of products or services provided by small groups of independent contractors are likely to grow. Organisations will therefore reduce in size.

Coupled with this is the increased use of new technology within the enterprise to run the enterprise. This too may lead to a reversion to market rather than hierarchy at a number of points, as was argued in Chapter 7. We would expect to find, therefore, that as the use of new technology extends, organisations will become smaller, with more complex internal organisations in some cases, and with greatly increased use of contracting.

The question we have not discussed is whether there will be more or less use of the peer group as a form of organisation. The conclusion I would draw from the above analysis is that some of those enterprises that have been sloughed off from larger enterprises might well adopt such an organisational form. This might be true of some of those enterprises resulting from management buy-outs, and might be true more often in situations where a large firm has contracted out the provision of a service which is now provided by a small group of professionals, such as consultancy on computing, or market research analysis, for example. However, even in these cases, and certainly in the case of larger firms, it is likely that those who have managed to obtain access to the capital will prefer to *employ* others to help them provide the service, or product, and control them hierarchically, rather than let them enter into partnership where they would share the profits.

Conflict will still therefore exist, and will occur across all four of the areas where control is exercised. People will continue to have different preferences for forms of work organisation and design of jobs. Negotiation will still take place. The fundamental purpose of this book has been to help those taking part in these negotiations to understand clearly the possibilities open to them.

References

Abrahamson, B. (1982), 'F. W. Taylor *vs.* the Peer Group: The Contradiction Between Hierarchy and Spontaneity in Organisation Theory', Arbetslivscentrum Working Paper, Stockholm.

Alchian, A. and H. Demsetz (1972), 'Production, Information Costs, and Economic Organisation,' *American Economic Review*, 62 (December): 777–95.

APEX (1979), *Office Technology: The Trade Union Response*, London: Association of Professional, Executive, Clerical, and Computer Staff.

APEX (1985), *Job Design and New Technology: APEX Guidelines*, London: Association of Professional, Executive, Clerical, and Computer Staff.

Armstrong, P. (1984), 'Work, Rest or Play? Changes in Time spent at Work', in P. K. Marstrand, (ed.), *New Technology and the Future of Work and Skills: Proceedings of Symposium Organised by Section X at the Annual Meeting of the British Association for the Advancement of Science, August 1983*, London: Frances Pinter.

Babbage, C. (1963), *On the Economy of Machinery and Manufactures*, London, 1832; reprint edition, New York, 1963.

Baldamus, W. J. (1961), *Efficiency and Effort*, London: Tavistock.

Barron, I. and Curnow, R. (1979), *The Future with Microelectronics*, London: Frances Pinter.

Battelle Institute (1978), *Der Arbeitsmarkt in Baden-Württemberg—Technologische Entwicklungen und ihre Auswirkungen auf Arbeitsplätze in den Bereichen Maschinenbau und Feinmechanik/optik*, Battelle.

Benson, I. and Lloyd, J. (1983), *New Technology and Industrial Change*, London: Routledge and Kegan Paul.

Benson, J. K. (ed.) (1977a), *Organisational Analysis: Critique and Innovation*, Beverly Hills: Sage Publications.

Benson, J. K. (1977b), 'Organisations: A Dialectical View', Administrative Science Quarterly, Vol. 22, No. 1, 21.

Birch, D. L. (1979), *'The Job Generation Process'*, MIT Program on Neighborhood and Regional Change, Cambridge, Mass.

Bjorn-Anderson, N. (1976), 'Organisational Aspects of Systems Design', *R & D*, Dec.

Blauner, R. (1964), *Alienation and Freedom*, Chicago: Chicago University Press.

Braverman, H. (1974), *Labor and Monopoly Capital: The Degradation of Work in the Twentieth Century*, Monthly Review Press.

Bright, J. R. (1958), *Automation and Management*, Boston: Harvard University Press.

Buchanan, D. A. and Boddy, D. (1983), *Organisations in the Computer Age: Technological Imperatives and Strategic Choice*, Aldershot: Gower.

Burns, T. and Stalker, G. M. (1961), *The Management of Innovation*, London: Tavistock Press.

CBI (1980), *Jobs—Facing the Future*, London, Confederation of British Industry.

Chandler, A. D. Jr (1977), *The Visible Hand: The Managerial Revolution in American Business*, Cambridge, Mass.: The Belknap Press of Harvard University Press.

Clegg, S. (1981), 'Organisation and Control', *Administrative Quarterly*, Vol. 26, 545–62.

Clegg, S. and Dunkerley, D. (1980), *Organisation, Class and Control*, London: Routledge and Kegan Paul.

Coase, R. H. (1937), 'The Nature of the Firm', *Economics NS* 4, 386–405; reprinted in G. J. Stigler and K. E. Boulding (eds.), *Readings in Price Theory*, Homewood, Ill.: Richard D. Irwin, Inc., 1952.

Cooley, M. (n.d.), *Architect or Bee*, Slough: Langley Technical Services.

Cooper, C. M. and Clark, J. A., (1982), *Employment, Economics and Technology*, Brighton: Wheatsheaf Books.

CSS (1981), *New Technology, Society, Employment and Skills*, London: Council for Science and Society.

Dahl, R. A. (1957), 'The Concept of Power', *Behavioural Science*, Vol. 2, 201–15.

Davis, L. E. (1979), 'Job Design: Historical Overview', in L. E. Davies and J. C. Taylor (eds.), *The Design of Jobs*, 2nd edn., Goodyear Publishing Inc.

Doeringer, P. and Piore, M. J. (1971), *Internal Labour Markets and Manpower Analysis*, Lexington, Mass.: D. C. Heath.

Edwardes, M. (1983), *Back from the Brink*, London: Collins.

Edwards, R. C. (1979), *Contested Terrain: The Transformation of the Workplace in the Twentieth Century*, London: Heinemann.

European Foundation (1984), *Technological Development in Banking and Insurance: The Impact on Customers and Employees: United Kingdom*, Dublin: European Foundation for the Improvement of Living and Working Conditions.

Forester, T. (ed.) (1980), *The Microelectronics Revolution*, Oxford: Blackwell.

Fox, A. (1971), *A Sociology of Industry*, London: Collier Macmillan.

Fox, A. (1974), *Beyond Contract: Work, Power and Trust Relations*, London: Faber and Faber.

Francis, A. (1983), 'The Social Effects of CAE in Britain', *Electronics and Power*, Jan.

Francis, A., Turk, J., and Willman P. (1983), *Power Efficiency and Institutions*, London: Heinemann.

Freeman, C. (1984), in 'Keynes or Kondratiev? How can We get back to Full Employment?', in P. K. Marstrand, (ed.), *New Technology and the Future of Work and Skills: Proceedings of Symposium Organised by Section X at the Annual Meeting of the British Association for the Advancement of Science, August 1983*, London: Frances Pinter.

Freeman, C., Clark, J., and Soete, L. (1982), *Unemployment and Technical Innovation*, London: Frances Pinter.

Friedman, A. (1977), *Industry and Labour*, London: Macmillan.

Friedrichs, G. and Schaff, A. (eds.) (1982), *Microelectronics and Society: For Better or for Worse*, Oxford: Pergamon.

Galbraith, J. K. (1967), *The New Industrial State*, Boston: Houghton Mifflin Company.

Galbraith, J. R. (1977), *Organisational Design*, Reading, Mass.: Addison-Wesley.

Gallie, D. (1978), *In Search of the New Working Class: Automation and Social Integration within the Capitalist Enterprise*, Cambridge: Cambridge University Press.

Gershuny, J. (1978), *After Industrial Society: The Emerging Self-service Economy*, London: Macmillan.

Goldthorpe, J. H., Lockwood, D., Bechhofer, F., and Platt, J. (1968), *The Affluent Worker: Industrial Attitudes and Behaviours*, Cambridge: Cambridge University Press.

Goldman, P. and Van Houten, D. R. (1977), 'Managerial Strategies and the Worker: A Marxist Analysis of Bureaucracy', in J. K. Benson (ed.), *Organisational Analysis: Critique and Innovation*, Beverly Hills: Sage Publications.

Goodman, P. S. (1979), *Assessing Organisational Change: The Rushton Quality Of Work Experiment*, New York: Wiley–Interscience.

Green, K., Coombes, R., and Holroyd, K. (1980), *The Effects of Microelectronic Technologies on Employment Prospects*, Aldershot: Gower.

Hackman, J. R. and Oldham, G. R. (1980), *Work Redesign*, Reading, Mass.: Addison-Wesley.

Handy, C. B. (1984), *The Future of Work: A Guide to a Changing Society*, Oxford: Blackwell.

Hatvany, J., Bjorke, O., Merchant, M. E., Semenkov, O. I., and Yoshikawa, H. (n.d.), 'Advanced Manufacturing Systems in Modern Society', working paper, Hungarian Academy of Sciences.

Heydebrand, W. (1977), 'Organisational Contradictions in Public Bureaucracies: Toward a Marxist Theory of Organisations', in

J. K. Benson (ed.), *Organisational Analysis: Critique and Innovation*, Beverly Hills: Sage Publications.

Holti, R. W. (forthcoming), 'The Nature of Control in Software Production', Ph.D. thesis, Imperial College, London.

Huws, U. (1984), *The New Homeworkers*, London: Low Pay Unit.

Hyman, R. (1972), *Strikes*, London: Fontana.

Ingham, G. (1970), *Size of Industrial Organisation and Worker Behaviour*, London: Cambridge University Press.

Ingham, J. K. (1974), *Strikes and Industrial Conflict*, London.

Institute for Research on Public Policy (1979), *The Impacts of Computer-communications in Canada: An Overview of Current OECD Debates*, Montreal. Nov.

IPM (1983), *How to Introduce New Technology: A Practical Guide for Managers*, London: Institute of Personnel Management.

Jenkins, C. and Sherman, B. (1979), *The Collapse of Work*, London: Eyre Methuen.

Kelly, J. (1985), 'Management's Redesign of Work: Labour Process, Labour Markets and Product Markets', in D. Knights, H. Willmott, and D. Collinson (eds.), *Job Redesign*, Aldershot: Gower.

Labour Research (1979), 'Microelectronics: The Trade Union Response', June.

Labour Research (1983), Special Issue on New Technology, Nov.

Labour Research (1984), 'Homeworking', July, 171–3.

Leontief, W. (1978), 'Observations on some Worldwide Economic Issues of the Coming Years', *Challenge*, Mar./Apr.

Leontief, W. and Duchin, F. (1983), *The Impacts of Automation on Employment: 1963–2000*, New York: Institute for Economic Analysis.

Littler, C. and Salaman, G. (1985), 'Beyond Bravermania', *Sociology*, 17.

McGuinness, T. (1983), 'Markets and Hierarchies: A Suitable Framework For an Evaluation of Organisational Change', in A. Francis, J. Turk, and P. Willman (eds.), *Power, Efficiency and Institutions*, London: Heinemann.

Mallet, S. (1969), *La Nouvelle Classe ouvrière*, Paris.

Marglin, S. (1974), 'What do Bosses do? The Origins and Functions of Hierarchy in Capitalist Production', *Review of Radical Political Economics*, Vol. 6, 2.

Marglin, S. (1984), 'Knowledge and Power', in F. H. Stephen (ed.), *Firms, Organisation and Labour*, London: Macmillan.

Marstrand, P. K. (ed.) (1984), *New Technology and the Future of Work and Skills: Proceedings of a Symposium Organised by Section X at the Annual Meeting of the British Association for the Advancement of Science, August 1983*, London: Frances Pinter.

Martin, R. (1981), *New Technology and Industrial Relations in Fleet Street*, Oxford: Clarendon Press.

Mintzberg, H. (1979), *The Structuring of Organisations*, Englewood Cliffs, New Jersey: Prentice Hall.

Mumford, E. (1983), 'Designing Secretaries', Working Paper, Manchester Business School.

Naville, P. (1963), *Vers l'automisme social*, Paris.

Nicholas, T. and Armstrong, P. (1976), *Workers Divided*, Glasgow: Fontana.

Nichols, T. and Beynon, H. (1977), *Living with Capitalism*, London: Routledge and Kegan Paul.

Noble, D. F. (1979), 'Social Choice in Machine Design: The Case of Automatically Controlled Machine Tools', in A. Zimbalist (ed.), *Case Studies on the Labor Process*, New York: Monthly Review Press.

Nora, S. and Minc, A. (1980), *The Computerisation of Society*, Cambridge, Mass.: MIT Press; first published in France, 1978.

Perrow, C. (1970), *Organisational Analysis: A Sociological View*, Belmont, Cal.: Wadsworth.

Pugh, D., Hickson, D., and Hinings, C. R. (1983), *Writers on Organisation*, Harmondsworth: Penguin.

Rothwell, R. and Zegveld, N., (1982), *Innovation and the Small and Medium-sized firm*, London: Frances Pinter.

Ruffieux, B. (1981), *New Information Technology and Employment*, EEC, DGV/A/2.

Sabel, C. F. (1982), *Work and Politics: The Division of Labour in Industry*, Cambridge: Cambridge University Press.

Simon, H. A. (1957), *Models of Man*, New York: John Wiley and Sons, Inc.

Simon, H. A. (1961), *Administrative Behavior*, 2nd edn., New York: Macmillan.

Sorge, A., Hartmann, G., Warner, M., and Nicholas, I. (1983), *Microelectronics and Manpower in Manufacturing*, Aldershot: Gower Publishing.

ST (1982a), *Programme of Data Policy for the Union of Civil Servants in Sweden*, Stockholm: Statst Janstemann Aforbundet.

ST (1982b), 'General Local Agreement Covering Direction and Basic Principles of the Computerisation of Administration Activities at the National Telecommunications Administration of Sweden', Stockholm: Stast Janstemann Aforbundet.

Stone, K. (1973), 'The Origins of Job Structure in the Steel Industry', *Radical America*, 7.

Stoneman, P., Blattner, N., and Pastre, O. (1981), *Information Technologies, Productivity and Employment: Analytical Study Based on*

National Reports, Paris: Organisation for Economic Co-operation and Development.

Swords-Isherwood, N. and Senker, P. (1980), *Microelectronics and the Engineering Industries: The Need For Skills,* London: Frances Pinter.

Thompson, J. (1967), *Organisations in Action,* New York: McGraw-Hill.

Trist, E. L. and Bamforth, K. W. (1951), 'Some Social and Psychological Consequences of the Longwall Method of Coal-getting', *Human Relations,* Vol. 4, 1.

Trist, E. L., Higgin, G. W., Murray, H., and Pollock, A. B., (1963), *Organisational Choice: Capabilities of Groups at the Coal Face Under Changing Technologies—The Loss, Rediscovery and Transformation of a Work Tradition,* London: Tavistock.

TUC (1979), *Employment and Technology,* London: Trades Union Congress.

Turner, A. N., and Lawrence P. R., (1965), *Industrial Jobs and the Worker* Boston: Harvard Graduate School of Business Administration, 1965.

Wainwright, J. and Francis, A. (1984), *Office Automation, Organisations, and the Nature of Work,* Aldershot: Gower.

Watts, A. G. (1983), *Education, Unemployment and the Future of Work,* Milton Keynes: Open University Press.

Weber, M. (1947), *The Theory of Social and Economic Organisation,* Glencoe, Ill.: Free Press.

West, J. (ed.) (1982), *Work, Women and the Labour Market,* London: Routledge and Kegan Paul.

Wilkinson, B. (1983), *The Shopfloor Politics of New Technology,* London: Heinemann Educational Books.

Williams, R. and Moseley, R. (1981), 'Trade Unions, and New Technology: An Overview of Technology Agreements', mimeo, Technology Policy Unit, University of Aston.

Williamson, O. E. (1975), *Markets and Hierarchies: Analysis and Antitrust Implications,* New York: Free Press.

Williamson, O. E. (1985), *The Economic Institutions of Capitalism: Firms, Markets, Relational Contracting,* New York: Free Press.

Willman, P. (1982), 'Opportunism and Labour Contracting: An Application of the Organisational Failures Framework', *Journal of Economic Behaviour and Organisation,* 3, 1, 83–98.

Willman, P. (1983), 'The Organisational Failures Framework and Industrial Sociology', in A. Francis, J. Turk, and P. Williams (eds.), *Power, Efficiency and Institutions,* London: Heinemann.

Willman, P. and Winch, G. with Francis, A. and Snell, M. (1984), *Innovation and Management Control: Labour Relations at BL Cars,* London: Cambridge University Press.

Wobbe-Ohlenburg, W. (1982), 'The Influence of Robots on Qualification and Strain', *Proceedings of Conference on Robotics in the Automative Industry, Birmingham, April 1982*.

Wobbe, W. (1979), 'Social and Organisational Aspects of the Introduction of Robots into Industrial Production', Working Paper V/79, Gottingen: Soziologisches Forschungsinstitut.

Wood, S. (ed.) (1982), *The Degradation of Work? Skill, Deskilling and the Labour Process*, London: Hutchinson.

Woodward, J. (1965), *Industrial Organisation: Theory and Practice*, London: Oxford University Press.

Woodward, J. (ed.) (1970), *Industrial Organization: Behaviour and Control*, London: Oxford University Press.

Work Research Unit, (1982), *Meeting the Challenge of Change: Case Studies*, London: Work Research Unit.

Index

Abrahamson, B. 117
ACTSS 164
agency theory 122, 151
Alchian, A. and H. Demsetz 122 n.
APEX 159, 162-3
Armstrong, P. 17
ASTMS 12, 159, 163, 166
atmosphere 111
AUEW (TASS), *see* TASS
authority relationship 123
automated guided vehicles (AGVs) 95-6

Babbage, C. 115, 127-8
Baldamus, W. J. 113
Barron, I. and R. Carnow 23
BBC 1, 10
Batelle Institute 11, 13
Benson, J. K. 72
Benson, I. and J. Lloyd 4 n
BIFU 165
Birch, D. 34-5
Bjorn-Andersen, N. 191
BL 31, 99-102
Blauner, R. 41-3, 46, 50
biscuit-making automation 81-4
bounded rationality 141
Bradford City Council 173
Braverman, H. 65, 70, 71, 72, 73, 106, 110, 115, 190, 201
Bright, J. 43-4
Brown, R. K. 113 n.
Buchanan, D. and D. Boddy, 81-4
Burns, T. and G. M. Stalker, 111, 145-8
Bylinsky, G. 1

CAD 2, 30, 32, 33, 135, 143, 148, 165, 169, 187-8
CAM 30, 148, 169, 187-8
Cambridge Economic Policy Group 12
capitalism
  influence on job design 29
  and the entrepeneur 128-9
CBI 177-8
Chandler, A. D. 140, 142
Chaplin, Charlie 38
class consciousness 45
Clegg, S. 72
  and D. Dunkerley 72, 117

clerical work and new technology 136-7
CNC 2, 11, 32, 69, 85, 89-93, 135, 143, 166, 188
Coase R. H. 141
Commons, J. R. 67
compensation theory 15-17, 198
  and its critique 17-23, 198
composite work roles 51-2
conflict at work
  as a general phenomenon 205
  Mallet's analysis 45
  Marxist perspective 41
contingency theory 44, 151
control
  changes in 137
  definitions of 45, 106
  four types 106-11
Cooley, M. 24, 25, 36
Cooper C. M. and J. A. Clark 21
co-ordination 44, 114, 120
Council for Science and Society 25, 29, 36
Courage's brewery 172
CPSA 163-4
craft workers 31

Dahl, R. 106
design and draughting jobs and task fragmentation 71
de-skilling 5, 7, 24, 38, 65-71, 72, 76, 86-8, 115, 157, 200
  and implications for organisations 71-8
direct control 77
division of labour 53
Doeringer, P. and M. Piore 124
dual labour markets 7

Edwardes, Sir Michael 178, 184
Edwards, R. 76, 77
EETPU 170
effort-reward bargain 109-11, 112, 122
employment effects of new technology
  and capital shortages 19
  changes in level 5, 9-23, 154
  changes in type 4, 9-10, 157

employment effects (cont.)
  due to productivity gains 13-4
  effects of new technology 10-23, 32, 33
  effects of unemployment benefit 18-19
  macro-economic modelling 14
  and need for job subsidies 20-1, 200
  predictions from firm-level analysis 13-14
  scenario building 14
  and slow economic growth 12-13
  theoretical considerations, *see* compensation theory
'Employment Society' 25, 26-7
ENIAC 2
entrepreneur
  definition 118
  role 125-8
EPOS 2
Equal Opportunities Commission 166
European Foundation 155

Financial Times 185-6
F. International 144, 166-8
Fleet Street 184-7
Forester, T. 1, 2, 12 n., 14 n.
Fox, A. 75, 111
fragmentation of jobs 5, 67-8, 72
Francis, A. 105 n., 120, 140
free-lancing 6
free-riding 109, 112, 122
Freeman, C. 22, 22 n., 27
frictional unemployment 18
Friedman, A. 76, 134 n., 158
Friedrichs, G. and A. Schaff 4

Galbraith, Jay 150-1
Galbraith, J. K. 139
Gallie, D. 45, 46, 49-50
Gershuny, J. 27-30
Goldman, P. and D. R. and Van Houten 72
Goldthorpe, J. H. *et al.* 40 n.
Goldwyn, E. 10, 11 n.
Gramsci, A. 117
Green, K. *et al.* 13

Hackman, J. R. and G. R. Oldham 38, 56-60
Handy, C. 14, 25, 26, 28, 30
Hatvany, J. 36
Henley Management College 89
Heydebrand, W. 72

high trust 75
Hoff, M. E. 1
Holti, D. 158
home-working, *see also* tele-commuting 166
Huws, V. 166
Hyman, R. 40 n.

ICL 167
Imperial College, London 148
implementation of new technology 171-196
industrial democracy 179
industrialisation and technological developments 61-5
Ingham, G. 40 n.
Institute for Research on Public Policy 11
Institute of Manpower Studies 13
Institute of Personnel Management 171-6, 195
Internal labour markets 73, 74, 77, 124, 158
International Institute of Management, Berlin 89
investment behaviour of firms 21

Jenkins, C. 12
job design
  criteria for well-designed jobs 56-8, 174
  design approaches 192-6
  experience at BL 182-4
  polarisation of skills 9, 24
  trade unions 7
job displacement effects of new technology, *see* employment effects of new technology
job rotation 87
job satisfaction and
  job design 56, 58, 60
  measurement problems 38-40
  sources of job satisfaction 40, 41
  and technical change 41
job subsidies and technological unemployment 20-1

Kalmar, Volvo Plant 93-7
Kelly, J. 174 n.
Kondratiev waves 22-3, 199

Laboratory for the Sociology of Work, Aix-en-Provence 89
leisure preference 17, 62, 116-17
'Leisure Society' 25-6

Lenk, K. 4
Leontieff, W. 15, 199
Leontieff, W. and F. Duchin 19, 21
Line-out system 98-9
Littler, C. and G. Salaman, 77
LO 168

machine pacing 101-2
Maddocks, I. 1
maintenance workers 31
Mallet, S. 45-7, 50, 60
managerial work 132-4
Manpower Services Commission (MSC) 164
Marglin, S. 62, 64, 110, 115-16, 120, 126, 128, 152
market *vs.* hierarchy 105-6
Martin, R. 185, 186
Marxist analysis 41, 48, 61, 113
McGuinness, T. 126
MDI (manual data input) 85
mechanistic organisation 11, 146-7
metering of effort 122-3
Metro, automated welding line for 31, 99-102, 178-184, 188
Mintzberg, H. 107, 120, 123
motivation and job design 56
multi-skilling 88
Mumford, E. 194

NALGO 165
Naville, P. 47-9, 60
new technology agreements (NTAs) 161-5
new technology
    impact on employment stability 34-6
    impact on size of firms 34-6, 137-44, 154
    the nature of work 24-30, 132-7, 154
NGA 160, 185
Nichols, T. and P. Armstrong 43
Nichols, T. and H. Beynon 49
Noble, D. 70
Nora, S. and A. Minc 11
Norway 36
Noyce, R. N. 2 n.
NUJ 159-60, 163
NUPE 16

occupational control 7
OECD 12
office automation 30, 34
opportunism 141

organic organisational form 146-7
organisational design 131-53
organisational types 117-25

part programming 135
payments systems 86, 156
Perrow, C. 149-50
polarisation of skills 92
power and control 106, 130
preferences *re* control 111-17
process automation 81-4, 85
producer co-operatives 117
professionalisation 5
professional work and new technology 134
proletarianisation 5
PTK 168
Pugh, D. *et al.* 118
putting out system 62

quality of working life movement 51, 56, 60

Rank Xerox 6
record-playback CNC programming 70
responsible autonomy 52, 53, 77, 158
robots 30, 33, 79-81, 165
Rolls Royce 177
Rosenbrock, H. 92
Royal London Mutual Insurance Society 177
Ruffieux, B. 13 n.

SAAB 51, 97-9, 101-2, 183
Sabel, C. 63
Sadler, P. 12 n.
SAF 168
Scandinavian experiences in new forms of work organisation 56, 93-7
Schumpeter, J. 22
Science Policy Research Unit 22, 27, 34, 35
scientific management 65, 73, 115
Scottish Provident Institution 176
secretarial work and new technology 136-7
Shell Petroleum 172
Sherman, B. 12
Siberia 36
Simon, H. 108, 140
skill shortages 31
socio-technical systems theory 51-6, 190
SOFI 33

Sorge, A. *et al.* 90-3
ST 169
stake-holders 111
Stone, K. 73, 115, 124, 127, 157
Stoneman, P. 13
Stonier, T. 11, 11 n., 14
sub-contracting 73, 76, 155-7
Swedish Co-determination legislation 103 n.
Swedish union experience 168-9
Swords-Isherwood, N. and P. Senker 30, 31, 32
system design 189-96

Tameside employment study 13
TASS 159, 163, 165
Tavistock Institute 51, 55, 59, 201
Taylor, F. W. 65, 66, 68, 72, 115
TCO 168-9
team-working 183
telecommuting 144-5, 156-7, 166, 202
TEMPO study 22
TGWU 166
Thamesdown Borough Council 173
Thompson, J. D. 111, 193
Times Newspapers 185, 186-7
trade unions and
  effects of new technology 114-15
  job design 7
  response to new technology 7, 154-70
transactions-cost analysis (TCA) 8, 122, 151

Trist, E. 51-5, 134 n.
Trist, E. and K. W. Bamforth 52, 158
TUC 132, 159-63, 170
Turner, A. N. and P. R. Lawrence 56

unemployment and new technology, *see* employment effects of new technology
'Unemployment Society' 25

Volvo 51, 93-7
Vonnegut, K. 26

Wainwright, J. and A. Francis 34, 189-96
Watts, T. 25, 36
Weber, M. 123
Weiner, N. 12
West, J. 191
Wilkinson, F. 85, 87, 88, 102
Williams, R. and R. Moseley 176
Williamson, O. E. 111, 118-19, 122 n., 123-5, 140-3, 203
Willman, P. *et al.* 99, 113, 125, 179-84
Wobbe-Ohlenburg, W. 79
Wood, S. 76, 77, 78
Woodward, J. 44-5, 48, 107, 148-9
Work Research Unit 163 n., 172
work organisation 7, 52, 55, 88, 89-93, 182-4
'Work Society' 25, 27